Sustainable Farm Finance

A Practical Guide for Broadacre Graziers

John C. H. Mitchell, Bruce J. Chapman
and David B. Lindenmayer

CSIRO

PUBLISHING

A catalogue record for this book is available from the National Library of Australia.

ISBN: 9781486316496 (pbk)
ISBN: 9781486316502 (epdf)
ISBN: 9781486316519 (epub)

How to cite:
Mitchell JCH, Chapman BJ, Lindenmayer DB (2022) *Sustainable Farm Finance: A Practical Guide for Broadacre Graziers*. CSIRO Publishing, Melbourne.

Published by:

CSIRO Publishing
Private Bag 10
Clayton South VIC 3169
Australia

Telephone: +61 3 9545 8400
Email: publishing.sales@csiro.au
Website: www.publish.csiro.au
Sign up to our email alerts: publish.csiro.au/ earlyalert

Front cover: (top) photo by Zoteva/Shutterstock.com; (bottom left to right) photos by Jed Owen/Unsplash, Africa Studio/Shutterstock.com, Finn Whelan/ Unsplash

Edited by Adrienne de Kretser, Righting Writing
Cover design by Cath Pirret
Typeset by Envisage Information Technology
Index by Max McMaster
Printed in Australia by Ligare

CSIRO Publishing publishes and distributes scientific, technical and health science books, magazines and journals from Australia to a worldwide audience and conducts these activities autonomously from the research activities of the Commonwealth Scientific and Industrial Research Organisation (CSIRO). The views expressed in this publication are those of the author(s) and do not necessarily represent those of, and should not be attributed to, the publisher or CSIRO. The copyright owner shall not be liable for technical or other errors or omissions contained herein. The reader/user accepts all risks and responsibility for losses, damages, costs and other consequences resulting directly or indirectly from using this information.

CSIRO acknowledges the Traditional Owners of the lands that we live and work on across Australia and pays its respect to Elders past and present. CSIRO recognises that Aboriginal and Torres Strait Islander peoples have made and will continue to make extraordinary contributions to all aspects of Australian life including culture, economy and science. CSIRO is committed to reconciliation and demonstrating respect for Indigenous knowledge and science. The use of Western science in this publication should not be interpreted as diminishing the knowledge of plants, animals and environment from Indigenous ecological knowledge systems.

This publication was supported by the Sustainable Farms project, The Australian National University.

SUSTAINABLE
FARMS

The paper this book is printed on is in accordance with the standards of the Forest Stewardship Council® and other controlled material. The FSC® promotes environmentally responsible, socially beneficial and economically viable management of the world's forests.

Contents

Disclaimer   iv

About the authors   v

Acknowledgements   vi

Glossary   vii

Preface   viii

Introduction   x

Chapter 1   **Taking stock and making plans**   **1**

Chapter 2   **Success begins with a budget**   **21**

Chapter 3   **Increasing your farm's profitability**   **38**

Chapter 4   **Financing farm expansion**   **59**

Chapter 5   **Achieving environmental sustainability**   **70**

Conclusion   78

Appendix 1: Your farm finance checklists   80

Appendix 2: Adverse-event plans   97

Appendix 3: Is a new property worth buying?   104

References   106

Index   109

Disclaimer

The advice in this book is general in nature and expert professional assistance should be sought in detailed financial planning and decision-making.

About the authors

John Mitchell has successfully farmed beef cattle and prime lambs for about 50 years. At the beginning of his farming career, after two years as a jackaroo, he completed a Bachelor of Arts in Economics and Accounting at The Australian National University and applied much of what he learned there to farm management. John is the lead author of this book, and he has personally experienced many of the trials that are discussed, including succession problems, drought, debt and bushfires.

Bruce Chapman is a labour-market and education economist who specialises in public policy related to government financing. He designed the Higher Education Contribution Scheme (HECS), the world's first national income-contingent loan for university students, which has since been adopted in 10 other countries. Bruce believes that income-contingent financing – where debts are repaid in the future when and (only) if borrowers have the resources to do so comfortably – could be applied effectively to farming. In that context, he is actively engaged in the ANU Sustainable Farms project.

David Lindenmayer has worked as a researcher on Australian farms for almost 25 years. He has a particular interest in improving environmental conditions on farm properties, including protecting remnant native vegetation as well as restoring and replanting it. He specialises in establishing and maintaining large-scale, long-term research and monitoring programs on farms. He and his field team have worked on ~250 farms from central Victoria to south-east Queensland for over 20 years.

Acknowledgements

John Mitchell thanks Professor David Lindenmayer for inspiring him to put this book together. He also thanks Professors A.H. Chisholm, R.S. Bird and A.D. Barton for their economics and accounting teaching at The Australian National University in the 1970s. He thanks the many farmers with whom he has exchanged ideas about agriculture and farm management over the years, particularly Don Lawson OAM and the late Michael Guinness.

Bruce Chapman thanks all the funders of and participants in the Sustainable Farms project at The Australian National University. He is particularly grateful to The ANU College of Business and Economics, which has contributed generously to the financial resources of the Sustainable Farms project.

David Lindenmayer thanks all the funders and participants involved with the Sustainable Farms project at The Australian National University. Key supporters of the project include the Ian Potter Foundation, the federal Department of Agriculture, Water and the Environment, the William Buckland Foundation and the Calvert-Jones Foundation.

We thank Pippa Carron, Louise Thurtell and Kathleen Weekley for extensive expert editing that enabled this book to be completed. We thank Dr Leo Dobes for comments and insights into the section on cost–benefit analysis.

Glossary

Abatement	Reduction in environmental and agricultural damage.
Amortisation	Payment of both principal and interest on a loan.
Arbitrage	The difference between two prices or payment rates (e.g. interest) which can lead to profit or loss.
Debt tunnel	The increasing level of debt on a farm because repayments on the original loan cannot be made.
Discount rate	The amount of income you expect to receive from a guaranteed investment (usually an interest rate); used to calculate a net present value or discounted cash-flow value.
Discounted cash flow	The value of an investment in today's terms taking into account expected future financial returns.
DSE	A 50 kg wether maintained at constant weight has a dry sheep equivalent (DSE) rating of 1. Animals requiring more feed have a higher DSE and animals requiring less feed have a lower DSE.
Ecosystem services	The many and varied benefits that humans gain from the natural environment and from properly functioning ecosystems.
Law of diminishing marginal returns	The decrease in the extra output gained when one factor of production is increased, but all the other factors remain constant.
Mean	The average of a list of numbers (the sum of all the numbers divided by the number of numbers).
Monopsony	A market largely controlled by one major purchaser of the goods and services offered by many sellers.
Net present value	The present value of a sum of money, in contrast to some future value when it has been invested at compound interest (calculated as the difference between the present value of cash inflows and outflows generated over time).
Return on capital	Net profit expressed as a percentage of net equity.
Riparian	Areas alongside waterways (creeks, rivers and dams).
Silviculture	The growing and cultivation of trees.
Soft commodity	Commodities that are grown rather than extracted or mined (e.g. beef, lamb, wool, grain, cotton).
Terms of trade	The ratio of a country's export prices to its import prices.
Time value of money	The greater benefit of receiving a specific sum of money now rather than later. Interest is paid or earned because it compensates the depositor or lender for the time value of money.

Preface

In 2017 we started a major project at The Australian National University called Sustainable Farms. It is an ambitious undertaking, spanning environmental, economic and mental health research, and is thus also a truly multi-disciplinary one, involving different academic colleges at our university but also incorporating markedly different fields and ways of thinking.

The very first donor to the Sustainable Farms project was John Mitchell. We didn't know his story at the time – which is one of immense difficulties and setbacks, but also extraordinary resilience and recovery. After the death of his father, John inherited responsibility for managing the family farm – and with it, its debts. Through good management and the application of his economic skills (as it happens, from his training at The Australian National University in the 1970s), John turned around the farm's financial problems. The division of assets just prior to his mother's death again created enormous financial challenges; yet again, John used his economic skills to solve the crisis. More recently, in the Black Summer fires of 2019–2020, John's farm, including most of the infrastructure, was razed. But once more, John solved the financial problems on his farm.

Although we – Bruce Chapman and David Lindenmayer – are academics, we have witnessed the financial hardship that has confronted so many people in agricultural Australia. Bruce has spent years of work on drought policy, for example, and David has personally witnessed bankruptcy in extended family farm businesses. It became apparent to us that John's way of thinking has important lessons for many other hard-working farmers in Australia. The practical ways he has used his economics and finance training to help solve the problems that life has thrown at him is a story that should be told to help others to limit the financial hardships that seem to be all too common for Australian farmers. This is why we believe that this project is so important. If, together, we can help even just a handful of farmers to better manage their farm finances, then the effort to help to produce this book will have more than justified the work to create it.

From here, this book is written from John's perspective to share his insights. The authors believe his first-person narrative voice will assist in the readability and accessibility of concepts that are usually presented in a dry and complex fashion. We hope you will agree.

David Lindenmayer and Bruce Chapman
January 2022

Introduction

In the nearly 50 years that I have been a farmer, I have dealt with droughts, bushfires and other tribulations inherent to Australian farming. The aim of this book is to pass on the knowledge I've gained in meeting these challenges, and to help you to make your farm both financially successful and environmentally sustainable. In the long run, financial success will be possible only when the farm is environmentally sustainable, and vice versa. Whether you are currently making a loss, breaking even, or simply believe you can do better, this book will help you to manage your farm finances more strategically, to return to profitability and even to increase it.

The crucial element in strategic management of farms is future planning, so planning is emphasised throughout the book. We especially look at how to approach the sometimes bewildering variety of decisions that you must make as a farmer – about expansion, investments, environmental improvements, labour recruitment, succession and so on. At each step, it is crucial to have as accurate a picture as possible of what you will be getting yourself into when you head down any of these roads. To gain the clearest picture, you will first need to 'take stock', which is where we start in Chapter 1. What sort of shape is your farm in? Is it viable? How financially fit are you and your farm to take on new projects or new directions? Is your farm worth increased investments of time and money, or should you sell before you head into deeper trouble? If you haven't done so already, you should probably master financial software and learn how to prepare applications for bank credit.

Most Australian farmers swing between good and bad years because of changing climatic conditions, variable crop returns, volatility in beef, wool and other markets, and natural disasters. This revenue instability, which can lead to an accumulation of debt, is why forward planning is vital to your farm operations. Chapter 1 discusses the importance of planning and advises you on developing three crucial plans: a business plan, an environmental plan, and an adverse-events plan. All of these are required if you are to make rational decisions and avoid returning to the bank again and again to borrow.

These plans are where you set your broad goals and projects, but you also need to draw up detailed budgets for the various aspects of your operations. In Chapter 2, you will find advice on drawing up budgets for maintenance, development, debt-reduction, the household, environmental improvements and other areas. We also discuss the problem of the 'debt tunnel', which is one of the most common traps for farmers – one that often leads not only to financial trouble but also to emotional difficulties, including intense stress and even depression, and can result in damaged family relationships. Sometimes it's hard to wait for the next cheque without borrowing again, but high levels of debt can become unsustainable and they erode the equity that you have in your farm. Monthly repayments on a large loan can impede your capacity to upgrade, and to buy better breeding stock or more efficient machinery. They also reduce your available cash to spend on environmental improvements such as reducing soil erosion, improving waterways and planting trees – all of which are crucial to the sustainability of your farm.

For all these reasons, in writing this book, I drew on the expertise of Professors Bruce Chapman and David Lindenmayer, who have economics training and experience in sustainable farming, respectively. Together, we've done cost–benefit analyses of various aspects of farming, such as extending land holdings and investing in machinery, to help you to make the best-informed decisions possible. In Chapter 2, we discuss the wisdom – for those farmers who have the financial capacity to do so – of reducing the concentration of risk in the farm by investing in off-farm assets such as residential or commercial real estate, or a share portfolio that pays dividends twice a year. Some farmers might also resort to part-time or seasonal work elsewhere. Other family members, including partners, might also work elsewhere during the week and then help around the farm on the weekend. Unfortunately, neither option leaves a lot of time for R&R for the farmer's family, but you knew that when you started farming!

Chapter 3 explains the importance of cost–benefit analyses and how to do them, in the context of a discussion about how you might improve your farm's profitability. Most farmers who want to expand their farming businesses decide to do so by either buying a neighbouring farm or shifting to intensive production methods. I understand why a farmer might want to do this, but in Chapter 3, I stress the importance, before taking either of these steps, of considering the law of diminishing returns (the theory that, at a certain point, the gains to be made from increasing one aspect of production will get smaller and smaller if all other aspects remain the same). The chapter deals with the kinds of questions you should consider before buying a neighbouring farm. Is the property worth buying? Can you afford the loan repayments? How can you optimise your operations without taking on more debt and more trouble down the track?

It is important for farmers to keep costs as low as possible if they are to increase profits. This may seem illogical at first, because lowering costs usually leads to decreased income, and yet it can lead to greater profit. For example, if total farm revenue is $100 000 per year and costs are $95 000, then a farmer is left with $5000 profit. Reducing costs to, say, $50 000 per year might result in lower revenues of, say, $70 000, but that would provide a farmer with $20 000 profit. For reasons explained in Chapter 4, reducing costs is also likely to be better for the environment, ecosystem condition and biodiversity. As farmers, we are only custodians of the land we work. We should aim to leave the infrastructure and the environment in better condition than when we took the farm over. Chapter 5 talks about how to do this by attending to pasture, weeds, soils, trees and erosion.

Good succession planning is particularly important for farmers, whose primary asset is usually the farm they own and work. Although succession planning can be confronting, and difficult to discuss with the family, it is imperative that you make your plans and have those discussions; everyone in the family should know what will happen when you retire, or if you die unexpectedly. Good succession planning reduces the likelihood of damaging disputes that can rupture family relationships and incur legal fees. This delicate topic is discussed in Chapter 1 and has its own checklist in Appendix 1E.

I have a passion for farming. I want to help Australian farmers to be more proactive, to improve their money management skills, to take control of their farm's finances and to think about improving the environmental aspects of their farm. Through years of experience, I have learned that financial security and prosperity are enhanced by prudent and proactive financial management. Having lived through good times and bad, drought and bushfires, high and low commodity prices, the formula I have arrived at for being a successful farmer involves some key activities. Each of these is discussed in detail in this book:

- developing a comprehensive farm plan that includes a business plan, an environmental plan, an adverse-event plan and a succession plan (Chapter 1)
- crafting a whole-farm budget that includes budgets for maintenance, development, debt reduction, family expenses and environmental improvement (Chapter 1)
- aiming for little or no farm debt and, if possible, making off-farm investments (Chapter 2)
- doing cost–benefit analyses of potential productivity improvements, taking into account economies of scale and the law of diminishing returns (Chapter 3)
- considering the pros and cons of buying more land (Chapter 4)
- learning and using environmentally sustainable farming practices (Chapter 5).

Appendix 1 contains a collection of Farm Finance Checklists, which you can refer to constantly as you consider all your options. I hope this book helps you to manage your farm finances in a more strategic manner and to return to profitability if you're currently making a loss, or to increase your profitability if you're breaking even, or simply believe you can do better. Whatever your situation – whether you have significant debt or not – I want to reassure you that most farms *can* be managed so they carry little or no debt. Most farmers *can* accumulate financial assets. And, assuming they're realistic, most farmers *can* have the level of prosperity they dream about.

There is money to be made in anything done properly, legally and ethically, and that includes farming. I hope that this book will help you on your way.

John Mitchell, BA, OAM
Towong Hill Station, January 2022

1

Taking stock and making plans

A family farm is more than just crops and animals. It's a way of life, it's heritage, it's the future. But it's also a way of life that ultimately requires the family farm to operate at a profit.

Whether you have inherited a farm or bought one,[1] most farmers have chosen farming as a way of life. However, you still need to run your business as a profitable enterprise. A significant number of family farms in Australia don't generate enough income to pay the family bills, let alone pay off debt. Farm debt in Australia has increased by over 75% in the last decade. But if you take a prudent and proactive approach to management, your farm can very likely be managed so that it carries little or no debt.

Given the variability of returns from year to year, it is crucial that farmers build a financial buffer to tide them over during difficult times. As well as trying to maintain some bank savings, you should consider using the federal government Farm Management Deposits Scheme,[2] which allows farmers to put pre-tax income into the scheme during good years and draw on it as needed during leaner years.

In 2018–2019, the average Australian farmer was 58 years old and had roughly 37 years of farming experience. The average equity ratio among broadacre farmers – the amount of debt divided by the value of the farm – was 88%. That is, the amount they owed to the bank was less than the value of the property. If you decide to sell your farm – or your succession plan involves selling the farm – then that equity ratio is a good sign. If, like many Australian farmers, you want to pass on the farm to a family member or members, then that level of equity is a good thing: if you and your partner don't have enough off-farm assets to generate a retirement income sufficient for you to retire in comfort and security, then your child(ren) will probably have to help you financially for the rest of your lives. A strong farm balance sheet (high equity) makes it easier for those who inherit the farm to pay such an annuity.

So, how to go about creating a successful farm? Is it feasible, given where you're starting from right now? To assess the potential of your farm to become a financial success in the long term, the first thing you need to do is take stock of your current financial situation, by compiling a simple net worth statement that adds your assets together and then subtracts your liabilities (explained below). Be warned: there is a risk that this exercise will make you see the difficulty of bringing your farm to long-term, sustainable profitability, and indicate that perhaps you would be better off selling rather than carrying on (at least with this particular farm). Despite the emotional connection you probably feel to farming and to your farm, it is sometimes better to get out early and preserve what equity you have – and at the same time preserve your mental health and your family relationships. Selling your farm might be the most appropriate move if you are in a debt tunnel, are making large losses after servicing the debt, and the debt is growing exponentially.

If, having filled out your net worth statement, you do decide that it is best to sell your farm, then you should sell well before foreclosure. You should put it up for sale only after completing fundamental tasks like repairing gates and fences and tidying up sheds and yards. Other relatively inexpensive things you can do to make the farm ready for sale and increase its sale price include:

- repairing and painting the house
- sprucing up the garden
- making any necessary repairs to farm buildings
- fixing any damaged sections of stockyards.

The mechanics of taking stock

There are numerous online resources to help you to compile your assets and liabilities statement, including an Agbiz calculator published by the Queensland Government (at the time of writing this was available from www.publications.qld.gov.au/dataset/agbiz-tools-business-and-finance-farm-finance/resource/090eebbc-3f41-4577-882f-a4694488106e) (see Figure 1.1).

One of the most important questions you must be able to answer is whether you will be able to afford the repayments on your various loans in both good years and bad. The value of your assets (land, stock, plant, licences) minus your total debts = your net worth. Your total debt amount, along with the bank's interest rate and the length of the loan term, determine the level of debt service (repayments). This

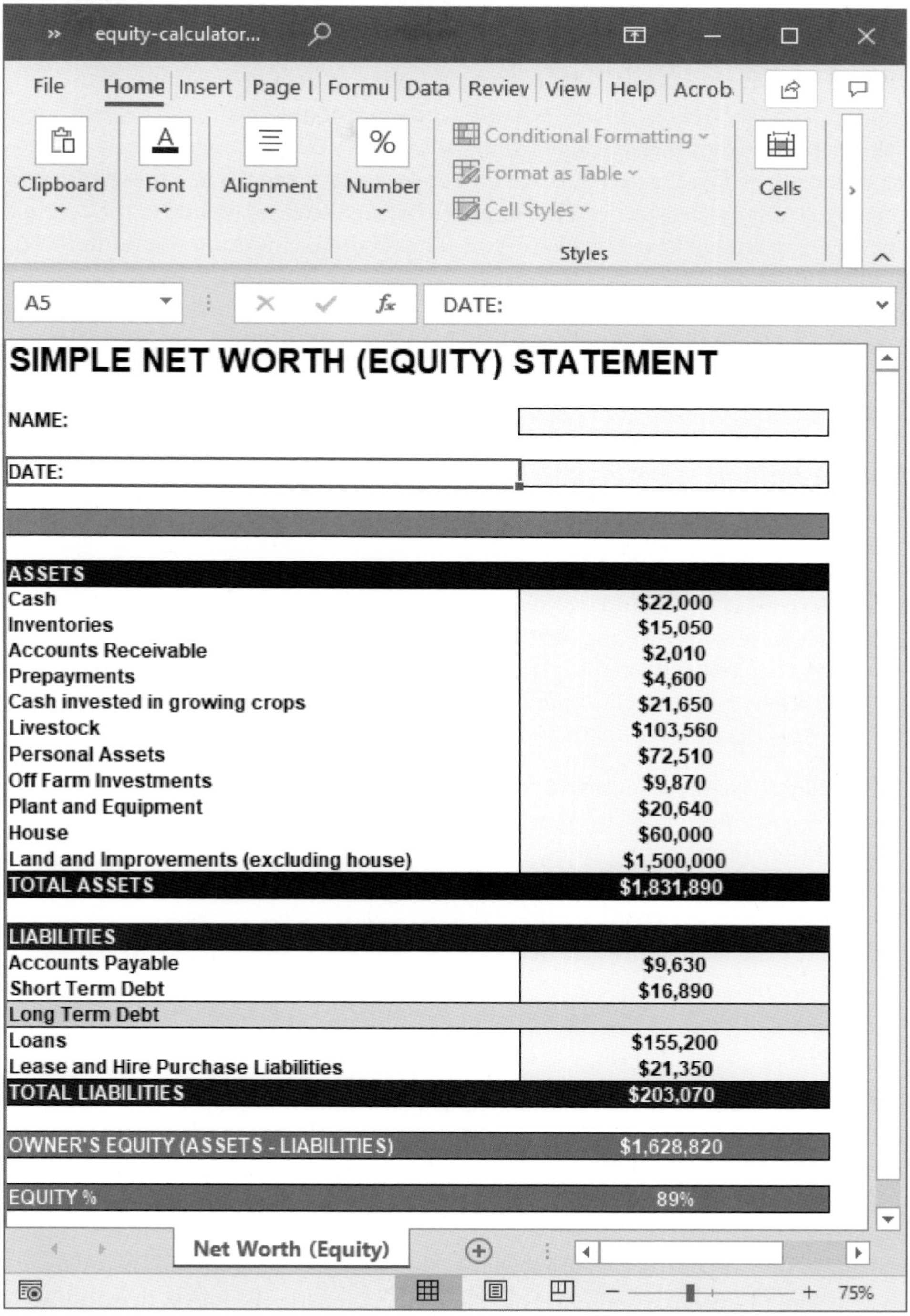

SIMPLE NET WORTH (EQUITY) STATEMENT

NAME:

DATE:

ASSETS

Cash	$22,000
Inventories	$15,050
Accounts Receivable	$2,010
Prepayments	$4,600
Cash invested in growing crops	$21,650
Livestock	$103,560
Personal Assets	$72,510
Off Farm Investments	$9,870
Plant and Equipment	$20,640
House	$60,000
Land and Improvements (excluding house)	$1,500,000
TOTAL ASSETS	**$1,831,890**

LIABILITIES

Accounts Payable	$9,630
Short Term Debt	$16,890
Long Term Debt	
Loans	$155,200
Lease and Hire Purchase Liabilities	$21,350
TOTAL LIABILITIES	**$203,070**

OWNER'S EQUITY (ASSETS - LIABILITIES)	$1,628,820

EQUITY %	89%

Net Worth (Equity)

Figure 1.1: Screenshot showing a simple equity statement

Queensland Government publications website also has a table which shows the yearly repayments that are required to service various amounts of debt. This is crucial knowledge – if your debts keep accumulating, then you might end up incurring penalty interest. There is a critical degree of indebtedness at which a farm with a certain turnover will head into what is called a debt tunnel, increasing the risk of mortgage foreclosure. You can end up in a debt tunnel when the debts owing on your farm grow because you are unable to make the repayments on the original loan. When payments can't be made, you can end up paying compound interest (i.e. paying interest on the existing interest) which adds yearly to the loan principal and thus the total amount of debt grows. This means that the proportion of your equity in the farm is declining (which can be worsened if land prices are not increasing, or are even falling, at the same time). The downward spiral could well end with the bank foreclosing on your mortgage.

Making plans

> Farmers don't plan to fail, they fail to plan.

Once you've taken stock and worked out your current financial situation, if you're keen to decrease your existing debt and to improve the financial performance of your farm, then the next step is planning. Logical and realistic planning is vital, and comprehensive farm plans are essential for many reasons. Farms plans are important because they:

- provide a baseline to measure financial success
- can be used to support applications for loans for expansion or other improvements
- help to integrate farm activities with environmental improvement and sustainability
- provide guidance in the face of adverse events such as droughts, fires, floods and cyclones
- provide a strategy for the succession of a family farm from one generation to the next.

A written plan that includes financial goals, completion dates, cash-flow budgets, environmental remediation and improvement strategies, and their timing, will go a long way to putting you on the path to financial success. Creating a plan will also give you a greater sense of control over your destiny and a feeling of personal achievement.

Failing to plan, on the other hand, can lead to financial ruin and bankruptcy. It can also lead to the environmental degradation of a farm, the needless suffering of stock animals and wildlife, and personal misery for you and your family.

Developing a prudent financial plan means budgeting for good and bad years. With a financial plan and sound budgets, you are less likely to fritter away income made in the prosperous years on desirable but non-essential tools, equipment and farm machinery. Instead, you will be more likely to put surplus money aside to accumulate for use in adverse times, or to reinvest in farm expansion, environmental remediation or improvement, and other financially productive activities.

Your farm plan should have four key components:

1. an overarching business plan
2. an environmental plan
3. an adverse event(s) plan
4. a succession plan.

The overarching business plan

The aim of your general business plan is to show if and when you will make a profit. It should include both big and small goals, then outline the steps you will take to achieve those goals within a projected timeframe, taking into account the costs of each of the four components. When I do a farm business plan, I follow the SMART maxim, which stands for:

- **S**pecific
- **M**easurable
- **A**chievable
- **R**ealistic
- **T**imely.

Your goals should include maintenance tasks and improvements to practices, stock, infrastructure and the land. Your planning period might be up to 25 years, with an initial focus on the first five years.

Try to quantify each project, setting out the labour and materials they will entail and their estimated costs. Projects might include things like removing noxious weeds, mitigating erosion, repairing farm buildings, rebuilding damaged stockyard sections, pulling down redundant buildings, making space for a decent and lockable workshop, repairing fences and gates, improving pasture, stock genetics and stock water supply, and starting a tree-planting program. Put dates against proposed expenditure to form a project timeline, and then budget for the various components and activities.

There are many useful business plan templates and guides available on websites maintained by a variety of government agencies, farming associations and commercial organisations. One of the best is ANZ New Zealand *Rural Tools and Templates* web page (which at the time of writing was available at www.anz. co.nz/rural/resources-insights/rural-tools-templates/). It allows you to download a farm business plan template as well as a clear and encouraging guide on how to fill it out.

The ANZ template is simple to use. It has boxes that you can fill in on your computer, or by hand on a printed copy. It is divided into the following sections:

1. business purpose
2. people
3. business environment
4. business assets
5. business performance and outlook
6. financial structure and gearing
7. options to increase performance
8. strategic position (SWOT)
9. plan (with a table divided into What, How, Who, When and Expected Results)
10. budgets, liquidity and profit
11. critical success factors
12. risks and issues
13. monitoring and performance management.

Whether you find such a task daunting or not, it is always worth asking people you trust or who are recommended by others for help in devising your plan. Such a person might be a farmer who is doing well and has a reputation for helping others, an accountant with agricultural experience, or people who represent or are recommended by your industry association.

There are many free online resources for farm planning, including the logistics and financial aspects.[3] There are also many commercial electronic farm management software systems[4] that have accounting capabilities, can forecast and track profits and losses, and have a range of other tools for planning and reporting, including digital calendars to schedule farming operations, track and measure field activities, livestock management, crop management, labour management, field-worker progress and risk management, and record costs for each application per head and per hectare. They can also generate comprehensive reports against key performance indicators.

Your business plan will have to be reviewed and updated regularly – at least quarterly but preferably monthly. If you leave it for too long, you might struggle to remember what you have already done and what new items you had planned to include. In addition, having an up-to-date plan, along with budget statements, is extremely useful when negotiating with banks for long-term and seasonal lending.

Along with the various plans and updates to plans, it is important to maintain a list and a photographic record of your achievements, so that you can review progress at a later date. As the months and years go by, it is heartening to be reminded of what your farm looked like before you started your improvements.

It doesn't matter how you start, even if it's just rough notes on an envelope; what matters is that you start! Your initial plan will be a long way from perfect, but remember that it is a living document, one you will change and add to repeatedly.

The environmental plan

Your overarching farm business plan should comprise other plans, as mentioned, including a budget for environmental improvements and biodiversity conservation. Not only is there intrinsic good in taking care of the environment, but it can also be done in a way that is profitable. Indeed, if you don't take care of the environment then your farm will become less profitable, because when a farm's economic and environmental health are allowed to deteriorate, the crop and wool yields per hectare will be lower and the weight of cattle and lambs will diminish, leading to a fall in profitability.

As with other plans, your environmental plan should include some goals and a strategy to achieve them. Your goals should include work in the following four categories.

1. Establishing several tree plantings with an understorey (undergrowth of shrubs etc.). You might, for example, plan on creating 10 tree plantings on a schedule of two plantings per year for five years. This will entail spending money on fencing, soil preparation and the purchase of tree and shrub seeds/seedlings.
2. Reducing erosion and restoring waterways. If you have creeks or rivers, this will involve removing weed trees and shrubs, planting native trees and shrubs, stabilising banks and installing protective fencing.
3. Controlling noxious weeds and pest animals in tree plantings and restoring riparian areas.
4. Improving soil, using techniques such as minimal tilling and direct drilling, as well as reducing or eliminating erosion using fencing, earthworks and plantings.

Consider including environmental improvement goals at a whole-of-farm level, that is, in a holistic environmental plan, then prioritise the most urgent tasks, such as limiting erosion around creeks or other waterways. You should prioritise tasks that will produce the greatest net financial benefit, such as pasture remediation.

As in all your planning, aim to balance urgent and easy tasks with work that might result in little benefit in the short term but that has significant benefits further down the track. For example, planting shelterbelt trees will have little immediate benefit because they take five to 10 years, sometimes even longer, to have an impact. However, in the long run they will be a crucial element in farm productivity by reducing soil erosion, protecting your flocks from adverse weather and providing habitat for native fauna.

You will have to consider the expense of different activities and do cost–benefit analyses that incorporate your total spending and the environmental gain per dollar. Cost–benefit analyses are discussed in detail in Chapter 3.

There is a wealth of online information about environmental planning, including how to establish and implement environmental projects on farms (and what kind of direct subsidies are available for tree-planting, fencing and other activities). The Meat and Livestock Australia (MLA) website, for example, has an excellent research and development section with a dedicated subsection on environmental sustainability (www.mla.com.au/research-and-development/Environment-sustainability). It provides useful and easy-to-understand information on how to run a sustainable grazing business, as well as practical tips on managing groundcover to reduce run-off and water loss, improving soil health and fertility, and grazing management. The website also has information on biodiversity and vegetation (www.mla.com.au/research-and-development/Environment-sustainability/biodiversity-and-vegetation), with practical toolkits for enhancing biodiversity. There are specifically targeted how-to modules for different types of graziers such as *Making more from sheep* (www.makingmorefromsheep.com.au).

Much helpful information about sustainable farming practices can be found on the Sustainable Farms website (www.sustainablefarms.org.au/) and the CSIRO website (www.csiro.au/en/Research/AF/Areas/Sustainable-farming-systems). If you still need to be convinced about the economic benefits of environmentally friendly farming, the internet offers many stories about farmers who improved their farm finances by investing in environmental sustainability works. Of special interest to broadacre graziers are some of the following examples:

- Farm 300, James Houston in the upper Murray region: www.youtube.com/watch?v=-_fwNlLfhoE&ab_channel=meatandlivestock
- Peter Holding in Harden, NSW: www.youtube.com/watch?v=anZ5Bg4TEr4&ab_channel=meatandlivestock
- Peter Whip from central Queensland: www.youtube.com/watch?v=VYBNdDFpC80&ab_channel=meatandlivestock
- The Mulloon Institute: https://themullooninstitute.org/

Other inspiring stories of farmers who use environmentally sustainable farming practices include those of banana growers in Queensland (www.youtube.com/watch?v=FFZnjI6wc6o) and dairy farmers in western Victoria (www.youtube.com/watch?v=2ft4US4M-uc).

Adverse-event plans

As a farmer, you have a special dread of bushfire, flood or cyclone destroying your farm, or a long and severe drought slowly killing your stock, trees, shrubs and pasture – not to mention challenging the livelihood and sanity of you and your family. Depending on where your farm is, other adverse events you might need to consider are landslides, hailstorms, unseasonal snow, plague locusts, mouse plagues and stock diseases. Some adverse events can result in a double financial hit – as well as destroying vital farming infrastructure that can be replaced relatively quickly, they can result in the loss of a major source of income by reducing or even eradicating stock, pasture and crops, which generally cannot be compensated for quickly. You have no control over the possibility of such a disaster, but if you write a proper adverse-event plan, then you will be better prepared for it. Farmers who are thoroughly prepared for natural disasters are less likely to suffer damage to property and loss of life. Their personal recovery time and resumption of farm production afterwards is usually faster, too.

Establishing a good relationship with your bank is important because it means that the bank will be more likely to help you during times of stress and financial hardship. Regardless of the strength of the relationship, however, if your farm is carrying substantial debt and is affected by an extreme weather event, then you will probably have to present a written financial recovery plan if you are to negotiate further credit or temporarily suspend interest and capital repayments. Ideally, the financial plan should include a two-year plan to rebuild the farm (prioritised point-by-point), the estimated cost of each task, who will do the work, a timeline for completion of the tasks, and a monthly cash-flow budget. In most cases, it will be worth paying an accountant for assistance to compile such a plan or at least to check your draft plan.

You should also consider a 'self-insurance' fund to help tide you over in the event of natural disasters. Options include off-farm investments such as shares or real estate (discussed further in Chapter 2), and a special savings account with your bank or the Australian Government Farm Management Deposits Scheme mentioned earlier, which allows farmers to put pre-tax income into the scheme during good years then draw on it later as needed (www.ato.gov.au/Business/ Primary-producers/In-detail/Farm-management-deposits-scheme/). These funds will help you to recover more quickly and easily, and with less stress. An indebted farm business with no external assets is at high risk of not financially surviving either a serious drought or sudden natural disaster.

For rapid-onset events such as fire, flood and cyclones, your plan should include three elements:

1. pre-event risk assessment and preparedness
2. actions to be taken immediately before and during the event
3. post-event recovery actions.

These components are illustrated in the example Fire Plan in Appendix 2A.

In addition to trying to establish an off-farm source of capital and income to ensure that you can survive rapid adverse events, you should:

- diversify your farm where possible. For example, don't have a very large farm of one crop in one area that is prone to fire, floods or cyclones[5]
- know where you stand financially and whether your business has the financial resources to replace lost infrastructure, stock and income until re-established
- have a written plan to respond to adverse events such as drought or bushfire, regularly review this plan, and add to preparedness whenever feasible
- track weather data and emergency warning websites on a regular basis
- act early as the event emerges
- have a personal and family survival plan, including early evacuation of your family, yourself and farm service animals and pets
- if you decide to stay and defend your property against a wildfire, choose a reliable, safe place such as a fire bunker, or install a sprinkler system
- stay positive and look for opportunities to enlist help from a variety of sources
- participate in, or keep abreast of, local community information programs
- join a cooperative that has an industry-specific recovery assistance plan[6]
- read survival stories of other farmers and devise strategies for rapid recovery.

Keep a copy of your plans off the farm, in both hard copy and electronic form. The latter can be on an external drive, in a cloud-storage service or on someone else's computer.

Since drought is endemic to many parts of Australia, the next section shows how to develop a plan for the possibility of drought. Appendix 2 offers some detailed adverse-event plans for rapid-onset events such as bushfire, flood and cyclone. There is also a useful checklist for 'Shock-proofing your farm' in Appendix 1C.

Developing a drought survival plan

Weather records for Australia show that, on average, a severe drought occurs somewhere in the country once every 18 years.[7] The southern half of the continent – where most broadacre grazing farms are – regularly experiences droughts of varying length and intensity. Few farmers in south-eastern Australia will forget the stress of the Millennium Drought, which started in 1999 and didn't break for 10 years.

Know your region

Unfortunately, and regardless of the scepticism of some media commentators and politicians, virtually all qualified climate scientists understand that the planet is warming as a result of human activities. There is also wide consensus among them that hotter, drier weather may well be a regular feature of Australia's future, resulting in more frequent and longer droughts. So, if your farm is in the southern part of the mainland, you must have a drought plan. Your first step should be to study your region's long-term weather patterns, to get an idea of the probability of a drought. While it is impossible to predict the onset and length of a drought, you can use historical rainfall and climate data to develop possible scenarios. The Rainman Streamflow software, for example, lets you work out local rainfall patterns based on 100 years of monthly and daily records.[8]

How much water and feed do you have?

Your second step should be to assess how well your farm could survive a drought, bearing in mind the number, size, depth and location of dams, as well as any creeks and rivers to which you have water access rights. If you're relatively new to the area, talking to neighbouring farmers will help you to make the assessment; compare weekly and monthly rainfall against long-term averages for any early indications that your region may be entering a dry period.

Use this knowledge to assess your need for stock feed and water, and how long you could expect feed and water to last at various times of the year. For example, open water will evaporate much faster in summer and breeding stock will need more protein resources than dry stock. Assess possible sources of least-cost feeding rations and donated fodder. Read the latest drought assistance information on the national government website (www.agriculture.gov.au/ag-farm-food/drought) and your state government site/s. NSW Local Land Services has a website page devoted to accessing assistance in times of drought or in the event of a natural disaster. The page also has a link to a form you can fill out should you need a transport subsidy for animal welfare. NSW also has a drought website (https://droughthub.nsw.gov.au/) described as 'Your one-stop online portal for NSW Primary Producer drought assistance and information'. It includes a financial assistance section, practical information for farmers in times of drought, and a section on wellbeing that lists the mental health services available to farmers. State website addresses change occasionally but you can always Google '[your state] drought assistance relief'.

Drought preparedness is a constant activity, and conservation of water resources both in your home and around the farm is crucial. Think seriously about whether you should be trying to increase your property's rainwater storage capacity by installing tanks to catch run-off from all shed roofs or adding more tanks if you already have some.

Your written drought plan should include:

- a water resource management strategy, including methods for tracking:
  › daily water use for farm buildings, stock and irrigation
  › total water storage capacity
  › monthly assessments of available water resources and soil moisture levels
  › identification of alternative water sources
- regular assessment of the state of infrastructure like pipes, troughs and pumps, to ensure that water wastage is minimised or eliminated entirely
- a stock management plan including what stock to sell and when as a drought progresses – to preserve pastures, you may need to sell stock early
- what extra feed you would need to have on hand, sources of supplementary feed, and the likely escalation of supplementary feed prices as a drought progresses
- what financial and business information and other assistance that state and federal governments, local councils and community organisations might provide in various circumstances
- how to maintain your mental resilience and emotional strength during a drought period, and a list of community and government resources to help you achieve this.

What are the financial risks?
If you have a bank loan, talk to your bank early in a drought to find out how much credit the bank will extend, and whether (and for how long) it will suspend your loan repayments.

A realistic assessment of financial risk and, in a worst-case scenario, an early decision to sell a farm may be the difference between selling at a profit or at a significant loss if the bank forecloses on your mortgage. You should talk to an accountant about your drought management options. If drought is declared in your area, you need to ensure that the Australian Taxation Office classifies your business as 'drought declared' and that you have decided that a forced sale of stock is necessary.

My own drought planning involves working out what fodder is on hand (approximate number of large bales of hay), stock numbers and the amount of feed in the various paddocks. These figures help me to determine the hay rations that the farm will need for the next 12 months. The hay rations affect the decisions that I make about which stock to sell and when. If there is little or no rain by a particular date, I have all my cows and ewes pregnancy tested, then sell all the dry cows and ewes because it is more lucrative to sell cattle and sheep while their weight is up rather than waiting till later when it may have dropped. If the drought continues, I sell aged cows and ewes, and keep the core of the breeding herd and flock for as long as possible. Calves are weaned early to give the cows a chance to hold their weight.

There is a river on my farm, so I initially graze hill paddocks that are at risk of running out of water, with a view to later moving stock into the paddocks that adjoin the river.

I also rework my cash-flow budgets and let the bank know that I might need to borrow a specific amount sometime in the near future. I show the bank manager my cash-flow budget and the dates scheduled for jobs to be done and decisions to be made based on rain (or lack of it) and available feed. This preparation gives me leverage with the bank because it shows that I am financially well organised.

My drought survival plan means that my decision-making is fairly automatic when a drought does occur. More importantly, it means that I don't stay awake at night stressing and worrying.

The time taken to write both a drought plan and an adverse-event plan is a crucial investment. It allows you to keep a core breeding herd and flock from which later you can breed up your stock numbers again, instead of having to borrow money to buy stock at highly inflated prices when the drought is over.

Insurance

I cannot stress strongly enough that, given Australia's relatively frequent droughts and other adverse weather events, it is vital to have up-to-date insurance policies for key areas of your farming business. Getting the right insurance can be complicated for farmers, but it is essential not only to cover yourself for when things go wrong, but also because most financiers will not lend against an item or property that is not fully insured. The various types of farm insurance policies that you should take out include:

- home and contents insurance
- farm motor vehicle insurance, which covers the additional risks to which your motor vehicles might be exposed because you work on a farm
- farm machinery insurance for tractors and associated equipment; if you lease your farm machinery, insurance is generally a mandatory part of the lease agreement

- farm liability insurance
- livestock, fencing and hay insurance to cover accidental injury, theft, damage, severe weather events and other mishaps to farm animals or produce
- loss of income and business interruption insurance for when accidental damage, theft, injury or adverse weather events prevent you from earning an income from your usual farm activities.

Other very important farm insurance policies include personal accident and injury insurance, goods-in-transit insurance, workers' compensation insurance if you employ people on your farm, and public liability insurance. Some insurance companies sell livestock insurance, but not many graziers take out these policies because they are so expensive. Insurance is often simply not available to cover crop losses, especially in flood- or cyclone-prone areas.

Insurers generally want to reimburse you as little as possible and take as long as possible to hand over whatever they agree to pay. Given this, if you make a claim against an insurance policy, you might need some professional help in standing your ground in the event of a protracted claim or pressure to accept less than you believe you are entitled to under the terms of your insurance policy.

While insurance will generally assist you with rebuilding infrastructure, it is unlikely to provide funds for environmental restoration, so dealing with this eventuality should be included in your recovery plans.

Many farmers are under-insured for buildings and other infrastructure such as yards and fences. If you can afford it, take out an insurance policy that will cover not only most or all of the cost of replacing buildings, equipment, infrastructure and stock but will also cover loss of income until you can reasonably expect your production systems to be up and running again, as well as accommodation costs should you lose your house in a fire, cyclone or flood.

Succession plans

A succession plan covers what happens to a farm business when the chief farmer can no longer run it, either voluntarily or through an accident. A will is not the only document you need for a complete succession plan; others include trust deeds and contracts, the farm plan, financial records and forward budget estimates.[9] You will find more details in the checklist 'Creating a farm business succession plan' in Appendix 1E.

Unfortunately, there is not a lot of succession advice available to farmers in regional Australia because there are few professionals with skills, knowledge and experience in this specialised area. However, there are some good online resources. The Farm Table website, for example, has a page devoted to succession planning including 'Dos' and 'Don'ts' and examples of successful succession planning. The

National Centre for Farmer Health also has a succession planning page with many links to useful resources.

You should have a serious succession plan by the age of 50. If you are older than that already, then now's the time to get started!

Ideally, you will have two succession plans: one in case you are badly injured or die unexpectedly, and another to cover what you want to happen when you retire as planned. A succession plan for death or incapacitation guards against the confusion and *ad hoc* decisions that can occur amid the turmoil and distress caused by serious ill-health or death. Financial damage, deterioration of the farm environment and stressed family relationships can all follow in the wake of a farmer's death or incapacitation. Retirement plans are just as vital as up-to-date wills because, with a well-considered retirement succession plan, you are more likely to be able to retire from the farm at a time and in a manner of your choosing, rather than in unhappy circumstances. Your retirement plan should cover the period from management succession to complete ownership takeover and involve full family agreement and documentation.

The key specification in both types of succession plan is documentation of what vital information exists (e.g. wills, loan and lease contracts, trust deeds, farm plans, budgets and tax accounts) and where the documents are stored. You should let a trusted family member or a trusted third party (a friend or lawyer) know where you have put these papers (they should be stored off-site, in case of fire). It is also important to choose competent and honest executors.

A family farm succession plan that is satisfactory to all parties takes time to finalise. It might take years to sort through the various options and the consequences of each option for individual family members. It can take even longer for each person involved to accept that your choices may not be what they were anticipating.

Many farmers want one of their children to take on their farm when they retire, die or are incapacitated, but sometimes all the children of a farmer want to pursue other careers and/or lack the ability and passion for farming. Sadly, if farmers try to force one of their children to take on the farm (e.g. by threatening to disinherit them), the result is often personally disastrous. It can also lead to an unproductive and unprofitable farm that is vulnerable to environmental degradation and ecological disrepair.

In many cases, the competing interests of a farmer's heirs result in the farm being divided up, resulting in the loss of economies of scale. Each parcel of land becomes less financially viable, leading to the possibility of overgrazing and overcropping. The knock-on effect is often a certain degree of poverty for the farming families who remain on the property.

Rural wills are often last-minute, brutal and susceptible to legal disputes. There can be disagreements between children and parents, and jockeying among the children for the inheritance. The consequences of this can be legal disputes. However, if you and your family have been genuinely involved in estate planning and can come to a mutual agreement as to the best future scenario, then it is much less likely that the financial assets will be depleted by legal and bank fees. If family members can't reach agreement, then it is wise to seek the assistance of a qualified estate planner and a mediation process to arrive at a fair outcome.

Sadly, many family farms have ended up in receivership as a result of little or no estate succession planning, especially when that lack of planning is combined with significant debt. The havoc can lead to considerable legal costs, financial destitution, badly fractured relationships, emotional damage to family members and, in extreme cases, suicide.

The damage to a farm's infrastructure and environmental condition caused when the farmer dies or retires while the farm is carrying severe debt can be enormous. When there is little or no labour or capital to put into physical maintenance or repair, sometimes for years, the condition of the farm deteriorates, sometimes rapidly, with increased weeds and feral animals among the problems. When the dispute is finally resolved, there may simply be no money left for environmental remediation by those who inherit all or some of the property.

A harrowing true story of succession disputes over a family farm

Bob and Elizabeth Smith were a salt-of-the-earth farming couple who owned a 4000 acre farm in NSW. Their daughters, Sarah and Judy, lived and worked in Sydney; their sons, David and Philip, worked on the farm and lived there with their respective families.

The farm made a reasonable turnover before interest, tax, depreciation and amortisation, but there was not enough income to support the three families living on it, especially while Bob and Elizabeth were supplementing the incomes of Sarah and Judy.

Both David's and Philip's wives complained about not having enough money for a half-decent lifestyle. David and Philip were good farmers and hard workers, but only had Year 10 education as Bob and Elizabeth didn't think that a Year 12 education was necessary for living on the land.

Both sons occasionally worked off-farm to supplement the family incomes. Their wives were homemakers, looking after their children, but had thought of trying to find other work later down the track. In Sydney, Sarah was single, working in administration and having trouble paying her rent. Judy and her husband had children and sometimes struggled to make their mortgage repayments.

The farm was originally in debt by $500 000. Interest was not serviced, so $80 000 was added to the debt each year for 10 years, at the end of which the debt had grown to $2 137 299. The loan calculations illustrating the financial problem of compound interest are shown in Appendix 3.

The value of the farm was $8 million, plus stock and plant of $2 million, making net assets of $7 682 700. So, Bob and Elizabeth were asset-rich but income-poor, with a big balance sheet by Australian standards. Unfortunately, this farm was in a debt tunnel, leaving Bob and Elizabeth in a very weak financial position.

Because of the chronic lack of surplus income from the farm, maintaining its infrastructure was difficult, as was making any farm improvements. Environmental works were simply unaffordable. The farm was overgrazed and there was no crop rotation, resulting in monoculture farming. Die-back in trees occurred because of a lack of remedial environmental work, leading to serious environmental degradation which was exacerbated as David cut down lots of trees to sell for firewood.

The daughters in Sydney became impatient: they were short of money and saw a large family asset that might help to relieve their situation. This is called 'early inheritance syndrome' (which in its strongest forms is regarded as financial elder abuse). But when Bob died, the farm was left to David and Philip. It turned out to be a poisoned chalice of an inheritance for them both.

The daughters Sarah and Judy were left out of Bob's will (sexism in the farming world means that a farm is often willed to the eldest son). They took Supreme Court action against the estate, which was probated at $7.7 million. Each daughter and son was represented by their own lawyer. Elizabeth, the widow, also retained a lawyer and yet another lawyer represented the estate – six lawyers in total, each with instructions to act against other members of the family and the estate.

The parties were not able to come to an agreement and legal costs reached $2 million.

With no legal resolution, the bank foreclosed on the farm, stock and equipment – all were sold in a fire-sale mortgagee auction. There were zero net proceeds after legal costs had been paid and the bank had taken $500 000 in foreclosure fees.

Family relations were destroyed. Elizabeth was left homeless and on the aged pension. Philip and David had to look for work, which proved difficult because of their lack of formal educational qualifications. Their marriages dissolved. David had a nervous breakdown from which he did not recover, and finally he committed suicide. Philip was left destitute and became an itinerant worker.

A 'tough love' recommendation

Too many Australia farming families fail to prepare adequately for farm succession, only to see whatever wealth they had vested in the farm whittled away in interest repayments and professional fees. But this doesn't have to happen. It is not always the case that a good succession plan will involve selling the farm; the farm can be passed on to the son or daughter who is running it while an equivalent value in off-farm assets is left to the other son/s or daughter/s not living on the farm. Or a second or third farm can be left to the other children, providing that each farm is of similar value, or is made up for in off-farm investments. However, often the only reasonable solution is to sell the farm and divide the assets. What follows is an adaptable generic recipe for a succession plan for a third-generation farming family comprising parents in their 60s with children in their 30s or 40s.

Let's assume that the farm (land, livestock and equipment) is valued at $4 million and there are three children. One daughter, Jane, is passionate about farming and has tertiary qualifications in farm management and 10 years' farming experience on the family farm and elsewhere. The other daughter, Anna, and the son, John, both live in a city, establishing their careers and trying to save for a house.

The succession arrangement for this family is simple though not easy, because it requires the sale of the farm. Selling would allow the parents to retire with $1 million, plus whatever other investment assets they might have accrued, and allow each child to receive $1 million. While it may be wrenching to see the farm pass to a new owner after being held for three generations, and cause emotional strain particularly for the parents and Jane (who had hoped to take over the farm), any grief about the sale will be considerably less than might be created by many other scenarios.

Jane uses her $1 million to buy some paddocks and equipment. She and her partner put their shoulders to the wheel and build up a new farming enterprise. The other two siblings put their $1 million towards buying houses in the city. The parents retire to a nearby town, with access to medical and health facilities. They look after Jane's children on a regular basis, and their other grandchildren from time to time. Family relationships are not fractured, there is no financial damage to any party and no substantial legal fees.

With luck, the family will sell the farm to new owners who have sufficient capital to invest in upgrading buildings and equipment as well as environmental remediation. Although this scenario doesn't follow the Australian rural tradition, that tradition has been known to take meals off tables and to cause tremendous financial, personal and environmental stress.

Summary of key points

- To run a profitable farming business, you need a comprehensive business plan.
- That plan should include (1) an overarching business plan; (2) an environmental plan; (3) an adverse-event plan; and (4) a succession plan.
- You can rough out your plan with pen and paper or on a computer. There are some great free online aids for farm planning, or you can buy sophisticated farm-management software that offers comprehensive activity tracking and reporting

tools. If you are not comfortable using such software, then consult an accountant if you can.

- The aim of your business plan is to show if and when you will make a profit or how you can increase your farm's profit. A key aspect of the plan is planning and budgeting for the various components of your farm operations.
- The aim of your environmental plan is to schedule landscape improvement activities at a whole-farm scale in the following four categories: tree plots; erosion control; noxious weed and vermin control; and soil improvement.
- The aim of your adverse-event plan is to ensure that you are adequately prepared for drought, fire, flood and cyclones at each of three stages: (1) pre-event risk assessment and preparedness; (2) taking action just before and during the event; and (3) post-event recovery.
- The aim of your succession plan is to have a documented strategy for withdrawing from your farm business, either as planned or unexpectedly, so that the transfer of your farm goes smoothly and your family avoids relationship breakdown and expensive legal and bank fees.
- You should set goals for all your plans following the SMART principles: Specific, Measurable, Achievable, Realistic and Timely. They should have costings and timeframes for achievement.
- You should update your plans frequently – at least once a year, and preferably quarterly or even monthly. Keep a list of achievements, with photographs, so you can review progress later and see the gains you've made.

The thoroughness and usefulness of your farm plans will be enhanced by online research and working with government, businesses and industry organisations to access relevant and up-to-date information. While I admit to being technologically challenged myself, I am assured by all who are not that there are many useful and user-friendly internet resources available for farmers.

Endnotes

1   If you don't own a farm but are thinking of buying one, then the checklist 'Buying your first farm' (Appendix 1A) will help you to decide whether it is the right move – and the right farm – for you.

2   ATO Farm Management Deposits Scheme: www.business.gov.au/Grants-and-Programs/Farm-Management-Deposits-Scheme#:~:text=The%20Farm%20Management%20Deposits%20(FMD,of%20their%20risk%2Dmanagement%20strategy.

3   For example, Dairy Australia provides assistance to dairy farmers through its 'Our Farm, Our Plan' initiative: https://www.dairyaustralia.com.au/farm-business/our-farm-our-plan.

4   For examples of farm planning and management software, see www.predictiveanalyticstoday.com/top-farm-management-software/. If you are going to buy specialised agricultural financial software, make sure that it can calculate net present value and discounted cash flow.

5   See, for example, banana-grower diversification designed to reduce risk at www.agrifutures. com.au/wp-content/uploads/publications/13-122.pdf.

6   For example, after Cyclone Larry in 2006 the Far North Queensland dairy industry formed a recovery team that supported farmers personally and financially. By the time that Cyclone Yasi struck five years later, the team had acquired collective experience which ensured that the industry was much better prepared and recovered much faster: www.agrifutures.com. au/wp-content/uploads/publications/13-122.pdf.

7   See, for example, www.leadingsheep.com.au/2012/04/drought-survial-stories/.

8   Rainman software is available at www.daf.qld.gov.au/business-priorities/agriculture/plants/ crops-pastures/broadacre-field-crops/cropping-efficiency/rainman.

9   More information can be found at Farm Table's 'Farm succession planning: an introduction and helpful guide': https://farmtable.com.au/farm-succession-planning-information/.

2

Success begins with a budget

Once you have a basic and viable farm plan, you should turn to setting budgets for the key areas of your farm activity, with the aim of achieving long-term sustainability.

The environmental, financial and emotional wellbeing of farmers is important both for you the farmer and for the farm itself. Farmers who improve the financial and environmental aspects of their farm are likely to feel more upbeat than those who do not. A farm business under financial stress is highly likely to also suffer environmental damage stemming from a loss of trees, poor soil health and limited biodiversity, all of which will reduce future output and decrease profit. A farmer's emotional health under these circumstances is likely to deteriorate, further compounding the problems. Together, these factors can lead in a downward spiral away from financial sustainability. Debt accumulating on the balance sheet quickly erodes equity and can ultimately mean that you will be forced to sell your property.

To put this scenario into the context of a farm balance sheet, let's look at a hypothetical example.

- Let's say that there is $800 000 worth of land and livestock and that the clearing sale value of the plant is $200 000, making total assets worth $1 000 000. Of this, $900 000 is equity and $100 000 is debt. Thus, the debt-to-equity ratio is 10%.
- Owing to financial hardship caused by a variety of factors, the debt is not serviced. On top of that, a further $10 000 is added each year because while the farm makes a profit of $30 000, annual living expenses total $40 000.
- The farmer has no money to spend on infrastructure maintenance or biodiversity improvements.

- Move forward 15 years and the debt has grown to $527 200,[1] a five-fold increase.
- The farm's total assets may have increased to, say, $1.2 million, less than might have been expected because the farmer has only been able to afford limited maintenance and the productivity and environment of the property have deteriorated.
- The farmer now has a debt of $527 200 and the equity is only $672 800, giving a debt-to-equity ratio of 43.9%, clearly in dangerous territory.
- The farmer has fallen victim to emotional depression and is incapable of recovery. This is a failed farm on all three levels – financial, environmental and emotional.

To prevent such a slide into debt and despair, it is important to have a comprehensive farm budget and to stick to it. Budgeting involves being very careful about expenditure. You need a realistic assessment of your financial progress each quarter and you must have an escape route at the ready. You need a solid understanding of at least three fundamental financial issues:

- the rate of return on your capital
- the size of your debt and its decelerating or accelerating future impact
- the benefit of savings and off-farm income, especially investment income with passive cash-flow.

Armed with this information, you can calculate your financial risk at any given moment, fine-tune your strategies to head off losses and be well on your way to financial success.

Financial success brings with it the capacity to improve the environmental condition of your farm, which leads to greater peace of mind. Budgets are *essential* to financial success.

How to be frugal

Farms can be run on either a frugal or a wasteful basis. If you're frugal, you think carefully about every purchase you make. When you're choosing a power drill, for example, which can cost from about $30 to $600, you should bear in mind that research shows that the average power drill is used for only 6–20 minutes in its entire life.[2] Prices of other tools also vary greatly and many of them will sit in your shed, gathering dust. Why not borrow or hire tools when you need them?

Many pieces of farm equipment quickly become obsolete or are used only once or twice. As with tools, you would do better to borrow or hire machinery that you'll need for only a short period.

If you must buy tools to have on hand, then try to get them at clearing sales, where many can be bought at a fraction of their retail cost. Not only does purchasing second-hand materials save a lot of money, it's also better for the environment.

Money saved by being frugal with tools and machinery can be put into Farm Management Deposits, invested in shares or put into environmental remediation projects where it will ultimately provide a good return to the farm in both financial and aesthetic terms.

Lock your sheds: tools and equipment stored in unlocked sheds are vulnerable to theft.

Cost–benefit analysis and budgeting are an integral part of farming. Both will help you to make the sort of decisions that will see you prosper rather than just break even or fail.

Creating budgets

Farm budgets should be divided into the following five categories:

- maintenance activities
- development activities
- debt reduction
- household expenditure
- environment remediation and betterment activities.

Use a separate spreadsheet for each budget, then add them together to create total figures in an aggregate cash-flow budget, which should match past and projected income and expenditure on a monthly basis.

These five budget subdivisions will allow you to accurately assess the value of various activities. For example, if you include fencing for tree plots and environmental exclusion areas in the general farm-works budget, then later on you will not be able to calculate the specific contribution of those environmental works to the farm's profitability. However, if you assign these costs to the environment remediation and betterment activities category, then you will be able to evaluate their specific benefits further down the track.

With your budget divided into these five categories, you will be able to assess where expenditure is too high and where you should cut back to achieve savings. Cost–benefit analyses will help you to make financial decisions for large items of expenditure. There are many business tools to help with these analyses. The Queensland Government, for example, offers a range of online agricultural business publications and resources you can use to calculate profits, construct budgets and cash-flows, and improve decision-making (www.publications.qld.gov. au/dataset/agbiz-tools-business-and-finance-farm-finance).

The following downloadable Excel spreadsheets can be found on that website:

- amortised loan calculator
- annual loan repayment equivalents
- equity calculator
- financial cash-flow budget
- financial options for plant purchases
- general farm accounting pack
- general farm budget template
- interest calculator for bills and loans
- lease calculator.

Maintenance budget

Your maintenance budget is for documenting routine farm expenditure such as stock, seed, fuel, fertiliser, stockfeed, animal husbandry products and procedures, and labour costs. It should include a contingency allowance for unexpected incidental expenses, such as veterinary treatments for livestock. You should be able to project this budget forward 12 months by adding in-line items of all expenditure from the previous 12 months, plus a 3% allowance for inflation.

Development budget

Your development budget covers essential upgrades to infrastructure and equipment, such as the cost of machinery replacement, new fences for further paddock subdivision, and laneways to increase the efficiency of stock management.

You can keep your development budget lower by not buying unnecessary new tools or machinery.

Debt-reduction budget

The servicing of loans is often the largest component of a farm budget. You should pay off your debts as quickly as possible. Consider increasing the amounts in your debt-reduction budget, and plan to reduce debt as much as you can in the profitable years.

Be mindful that although interest-only loans save money on a yearly basis and may be essential in lean years, they can be a financial trap in the long term. They certainly can create serious obstacles to a smooth business succession if you intend to hand over or leave your farm to your children.

Think hard before adding to your debt by borrowing for large items of equipment or infrastructure. You should plan to buy essential items of equipment and machinery using cash from the previous year's profit – if you can't do that, you should consider whether you really need the item and whether you can afford it in the light of your whole-farm budget and long-term financial goals.

Household budget

Your household budget should include all personal expenses (transport, food, clothes, medical, educational, etc.) for all the members of your family who live on the farm. Classify each item as 'essential' or 'nice to have', then cut out at least half of the 'nice to have' items. Analyse how the 'essential' costs can be reduced, even if only by a small amount. For example, a vegetable garden and slaughterhouse can considerably reduce your family food costs. A family of four would be doing well if they can keep their annual budget to under $40 000 (unless the children need to be sent to boarding school because the property is remote).

The General Farm Budget template on the Queensland Government's Agbiz tools, Business and Finance, Farm Finance website (www.publications.qld.gov.au/dataset/agbiz-tools-business-and-finance-farm-finance/resource/49a04dc8-3f32-49ab-b952-9289b4098113) is a downloadable, easy-to-use Excel document (see Figure 2.1). You can use it to input monthly costs in a slightly different way from the way I've outlined here, but it involves inputting the same figures.

Example budget for environmental works

When budgeting for the creation of 10 tree plantings (two per year) on an 800 ha property (one planting per 80 ha), the absolute minimum effective size for each planting is ~150 m × 40 m, or 0.6 ha. All 10 plantings together would thus comprise less than 1% of the total farm area.

To minimise costs, you should establish plantings along existing fence-lines wherever possible, making a total distance to fence of 230 m instead of 380 m. The cost of fencing materials (wire, intermediate posts and strainer assemblies) would be in the vicinity of $500 per 100 m for cattle and $850 for sheep,[3] giving a total of $1150 and $1955, respectively, for each tree planting.

Advice on soil preparation, tree selection, planting methods[4] and seed and tubestock costs can be obtained from Landcare or Greening Australia, or from books (see References at the end of this book). If possible, create an environmental corridor by connecting plantings or locating them near fenced riparian areas or environmental zones on adjoining properties. ANU research staff have written books advising about how wide the plantings should be (typically no narrower than four tree widths wherever possible).[5]

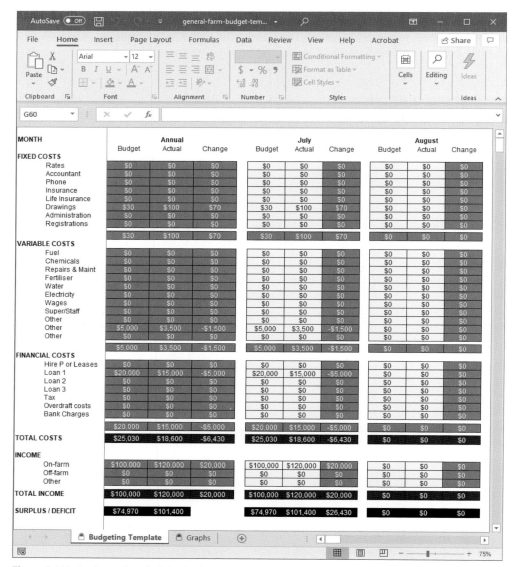

Figure 2.1: A simple equity calculator for farm assets

Environment improvement budget

Your budget for environmental remediation should include, at a minimum, costs for the establishment of tree plantings and the restoration of waterways (e.g. bank stabilisation, removal of willows, planting of native trees and shrubs). You should also include the dates by when you want to achieve each task and put those dates into your yearly activities schedule.

Aggregate cash-flow budget

A cash-flow budget is essential for farm planning, but surprisingly few farmers maintain one. You can create one using free or paid online resources or specialised accounting software,[6] but even jotting numbers down (in pencil, so that the figures can be changed later) in a ruled-up column book is better than nothing. Keeping an up-to-date cash-flow statement will help you to make better farm management decisions and to develop a stronger farm business. An aggregate cash-flow budget is vital when negotiating bank loans of a fixed nature and for supplementary or seasonal finance. When you have established the habit of maintaining a month-by-month budget, you will always know well in advance whether you are going to have a cash shortfall in any given period and be able to make plans for it.

If you've just sold a large number of sheep or cattle, for example, or bulk grain, then a cash-flow budget shows what portion of that cash you will need to put aside to run the farm and the household for the next 12 months, and how much of it you can put away for later when it will be needed. If you don't have a budget and you spend most or all of a large cash receipt on a vehicle or machinery, then you may be unable to pay the bills in three months' time. This will lead to cash-flow problems and a consequent loss of the bank's confidence in your ability to manage money.

Preparation of a reasonably accurate cash-flow budget removes much of the stress of farm decision-making and should eliminate costly impromptu expenditure decisions.

To create a simple cash-flow budget, first add together the total amounts in each of your five separate budgets (maintenance, development, debt reduction, family expenses and environment). With that, you have tallied expected expenditure and expected income for the year ahead, and can create a projected bank balance for any given month.

Your expected expenditure and income can mostly be calculated from past receipts, adding 3% for inflation. This might seem a lot at present,[7] but it will build in a contingency allowance for unexpected expenditure or increases in costs.

A cash-flow budget will show you when to expect a cash shortfall in any given period and thus when you will need to adjust your farm plan to deal with the likely deficit. Your spreadsheet should be updated at least monthly, and preferably fortnightly or even weekly: check expenditure and income against your bank balance. Updating the budget is important because it shows how accurate your projections have been. Comparing budgeted figures with actual results and analysing the reasons for any difference can be a highly educational exercise that will help you to create a more accurate budget in the future. *The more often you do this, the easier it will become to do, the less time you will spend on it in the long run, and the more control you will feel you have over your finances.*

When your cash-flow budget is done, stress-test your farm's financial resilience by introducing a fixed percentage drop in anticipated income and a fixed percentage rise in expenses to see how much these changes affect the bottom line – your bank balance. Run several scenarios – 2% drop, then 5%, 10% and 25% drops – and work out at what point it becomes financially scary. Think about what you would do if you got to this point.

Table 2.1 shows a hypothetical but realistic example of a whole-farm cash-flow budget. It reveals that annual expenditure was $4000 more than income over the year, which means that when the bank balance is brought forward from the previous year, the total Figure goes from $25 000 to $21 000 – a negative cash-flow that should be avoided. In this example, there are three periods of cash shortfall: May, August and November. A farmer seeing this projection would have to find ways to reduce expenditure by $4000 over the year just to break even, and reduce it by $10 000 in the first quarter to avoid having to ask the bank for an overdraft (which will entail extra interest payments). Alternatively, the farmer would have to find ways to increase income in the first three months of the year.

List of assets and liabilities

Keeping an up-to-date list of your farm's assets and liabilities helps you to make important decisions and assessments such as:

1. what stock to sell in the event of a drought
2. what your losses have been in the event of a catastrophe
3. approaching your bank manager about an overdraft
4. updating your annual insurance
5. whether you are in a position to purchase on-farm or off-farm assets.

Include an estimate of 'useful life' for each asset to identify whether they are 'current' or 'long-term'. Always assign conservative values to your stock and crops.

As mentioned in Chapter 1, there are free online tools to help you to compile your list of assets and liabilities. The assets list should include:

- cash and savings in long-terms deposits such as the Farm Management Deposits Scheme
- pre-paid expenses such as insurance
- accounts receivable, e.g. any moneys owed by the grain board or stock agents
- livestock (their saleyard value)
- stored hay, grain and stockfeed
- unharvested crops
- stored crops
- seed
- fertiliser
- infrastructure (including all residential buildings, sheds, tanks, fences, yards)

Table 2.1. Hypothetical but realistic cash-flow spreadsheet

Month	Jan	Feb	Mar	Apr	May	Jun	Jul	Aug	Sep	Oct	Nov	Dec	Total	Monthly average
Estimated income (inflows) ($)														
Stock sales			60 000						60 000				120 000	10 000
Crop sales						50 000						50 000	100 000	8333
Miscellaneous sales	3500		2000	1000			2500			1000	1000		11 000	958
Rental income	2500	2500	2500	2500	2500	2500	2500	2500	2500	2500	2500	2500	30 000	2500
Dividends		16 000						16 000					32 000	2667
TOTAL	6000	18 500	64 500	3500	2500	52 500	5000	18 500	62 500	3500	4000	52 500	293 000	24 458
Estimated expenses (outflows) ($)														
Maintenance	6000	6000	6000	6000	6000	6000	6000	6000	6000	6000	6000	6000	72 000	6000
Development			10 000						10 000				20 000	1667
Debt reduction	12 000	12 000	12 000	12 000	12 000	12 000	12 000	12 000	12 000	12 000	12 000	12 000	144 000	12 000
Household	4000	4000	4000	4000	4000	4000	4000	4000	4000	4000	4000	4000	48 000	4000
Environment				3000						3000			6000	500
Tax instalment			7000										7000	583
TOTAL	22 000	22 000	39 000	25 000	22 000	22 000	22 000	22 000	32 000	25 000	22 000	22 000	297 000	24 750
Surplus/deficit	-16 000	-3500	25 500	-21 500	-19 500	30 500	-17 000	-3500	30 500	-21 500	-18 500	30 500	-4000	
Cash in bank (brought forward)	25 000													
Bank balance (end of month)	9000	5500	31 000	9500	10 000	20 500	3500	0	30 500	9000	-9500	21 000		

- tools and equipment
- vehicles
- office equipment
- environmental works such as dams and spillways
- life insurance benefits
- off-farm investments such as shares and real estate.

Your list of liabilities should include:

- accounts payable
- credit card balances
- school fees
- vehicle lease payments
- operating lines of credit
- principal and interest due on loans.

Knowing the return on your capital

All businesspeople who invest capital in a financial enterprise seek a reasonable return on their investment. Farmers are no exception. You should know what the rate of return is on your farm for any given year. Knowing your annual rate of return on capital lets you compare it with other years to determine whether your business is becoming more profitable or less profitable.

This calculation also allows you to compare your farm with other forms of investment. If, down the track, you decide to buy another farm, understanding how to calculate the rate of return on capital will allow you to assess the potential purchase in a financially objective manner. Alternatively, if you decide to sell your current property, the annual rate of return will be the first thing an informed prospective buyer will want to know. You can calculate the rate of return on capital for your farm as follows.

A. Take your annual net profit, which is the adjusted net income on your tax return, calculated as follows:

- farm sales minus expenses and variations (up and down) of livestock numbers × market value plus fringe benefits (house rent-free and private use of phone and vehicle) plus depreciation on infrastructure and machinery, minus your wages as owner-operator, and plus or minus the change in the livestock inventory.

B. Take your net equity, which is the market value of what you own, calculated as follows:

- the total estimated value of your farm real estate plus the clearing sale value of machinery, plus the market value of your livestock, plus the market value of any irrigation licences, minus your total debt.

Table 2.2. Calculating the rate of return on capital

A. Net profit	$	$	
Farm sales	180 000		
Fringe benefits	30 000		
Expenses		80 000	
Owner wages		60 000	
Livestock inventory change		30 000	
	240 000	170 000	
Net profit (adjusted)			**$40 000**
B. Net equity			
Value of farm	2 050 000		
Sale value machinery	220 000		
Irrigation licences	100 000		
Total debt		800 000	
	2 370 000	800 000	
Net equity			**$1 570 000**
Return on capital = (70 000/1 570 000) × 100			**2.6%**

Annual fluctuations in farm income, and therefore in the returns on capital, are usually significant compared with the fluctuations in income from many other economic activities; some years can be very profitable and other years will return losses. Over 10 years, the average rate of return on a grazing property in south-eastern Australia, for example, is roughly 1.5–2.0%, which is well below the average return on five-year, risk-free, government bonds which are currently around 3.6%.8 This example showing 2.6% for a given year would fall below the non-farm average range. An average return over five years of less than 1% should ring alarm bells.

Calculate **A** as a percentage of **B** to get the rate of your farm's return on capital. Table 2.2 shows a hypothetical example.

The value of a farm as an investment

Given the financial risks involved in farming and the likelihood of going into debt after a natural disaster or a sharp reduction in commodity prices, we should add another four percentage points to allow for variation in income, and three points to represent the economic value of the skills, dedication and hard work required of a principal farmer. So, a total of 10.75% is the rate of return that farmers *ought to* receive from a farm. For the example shown in Table 2.2, this would require a mean net profit for the last 10 years of, say, $150 000, including fringe benefits; that amount is what they *would* receive from an asset of the same economic value in other industries. Excluding the farmer's salary, as above, the farm's economic value is thus $150 000 divided by 0.1075 = $1 395 348.

This is the economic value even though the market value of the enterprise may well be around $4–5 million, including livestock and equipment. In short, there's a large disparity between expected and actual return on investment, making a farm purchase prohibitive for many, and a poor investment for those already holding land. This is a major reason why some children of farmers do not want to inherit the farm but would prefer instead that the property be sold and they be given a share of the money to invest in other assets or enterprises.

Debt tunnels

If you are already in debt, or are considering going into debt to purchase land or make other investments, then it's a good idea to project the debt into the future to see whether it will ultimately improve your financial position or lead you down a debt tunnel.

Table 2.3 illustrates how borrowing $750 000 could lead a farmer who owns a 1000 ha property valued at $5 million down a debt tunnel. The servicing cost of this loan, at 4% interest, would be $86 025 per annum. If we add maintenance expenses of $25 000 and living expenses of $50 000, that makes a total of $161 000 per year which must be generated before any of the principal is paid off the debt. This income is simply beyond the commercial return of most 1000 ha grazing properties, year after year. A small number of mixed-crop farms might be able to make this amount, but not many.

If the loan is left unserviced (no payments are made on either interest or principal), after 10 years it will have ballooned to $1 619 175 and the farmer risks foreclosure. Keep in mind that this is a hypothetical example; in reality, the bank is likely to foreclose in two to three years, not 10 years. In the meantime, the ecological health of the farm is likely to have substantially deteriorated, reducing both its value and the farmer's equity. This is likely to affect the emotional wellbeing of the farmer and immediate family members. This example assumes a 1000 ha grazing farm in the NSW Southern Tablelands; a larger farm could support more debt.

The power of savings

You have a greater chance of gaining long-term financial sustainability if you build a buffer of savings to tide you over during times of unexpected financial stress such as when crops fail, or fire or floods occur. You should get into the habit of putting money into a savings account whenever possible, even if it's only $100 at a time. The amount that you deposit is less important than your action of starting a

Table 2.3. Result of servicing farm debt over 15 years at 6% and 8% interest

| Total debt ($) | Debt service per annum over 15 years | | Debt left unserviced after 10 years at 8% ($) |
	6% interest pa	8% interest pa	
250 000	25 325	28 625	539 725
500 000	50 650	57 350	1 079 450
750 000	75 925	86 025	1 619 175
1 000 000	101 300	114 700	2 158 900
1 250 000	126 625	143 375	2 698 625
1 500 000	151 950	172 050	3 238 250

Note: Debt service ratio at 6% = 0.1013 and at 8% = 0.1147.
Source: Table constructed from Pyhrr SA and Cooper JR (1982) *Real Estate Investment: Strategy, Analysis, Decisions*. Wiley, Boston, p. 777.

separate savings account; the positive psychological significance of saving is undeniable. Aim to make higher deposits during times of prosperity and resist the temptation to raid your savings for non-essential items. One of the best ways to do that is to put the money in a fixed-term deposit with a (slightly) higher interest rate as that makes it much harder to access the money quickly.

If you receive a large amount of money for the sale of crops or livestock, then put whatever you can spare into the Farm Management Deposits Scheme (www.business.gov.au/Grants-and-Programs/Farm-Management-Deposits-Scheme). Under this scheme, money deposited for 12 months or longer is not taxed until it is taken out, meaning that you can reduce your tax in a year of surplus, then take the money out in a year of low income. This is an effective form of risk management designed for Australian farmers whose annual revenues are highly variable.

Off-farm income

Given the variability in farmers' annual income, it is advisable, where possible, to have a source of non-farm income, whether that be off-farm employment or investments. Although it lacks the tax advantage of Farm Management Deposits, off-farm income through employment and investments can greatly reduce variation in income for a farming household.

Say, for example, that a farmer has an average annual taxable income of about $80 000. And say that the farmer invests $1 million in off-farm assets (a rental property) and/or fixed-interest financial products such as term deposits or government bonds. The market value and income of these investments have no connection to the farm revenues. The mean annual income of a portfolio like this would be $50 000 (5% of the investment outlay), and would vary much less than share prices and decidedly less than farm income. In total, $80 000 farm income plus $50 000 investment income would mean an expected annual average income of $130 000.

During drought periods, farm income drops by two standard deviations, which in this case would mean a fall to zero. Thus, with the $50 000 in off-farm income, the farmer would make just 39% of expected income. But without those off-farm investments, there might be no revenue for the year at all. So, off-farm investment means less variation in farm household income, resulting in less stress and making it easier for you to plan confidently for the future of the farm.

Off-farm investments are also handy for farm succession. In the scenario described in Chapter 1, where the farm was valued and sold at $4 million, providing $1 million for each child, the farming-oriented daughter, Jane, could take the farm, and the Sydney-based son and daughter could split the investment portfolio on the death of their parents.

It is sobering to realise that the expected rate of return on a farm business is only around 1.5% pa, which is much less than the return on the mixed real estate and investment portfolio at 5% pa mentioned above. This reality confirms the wisdom of budgeting and saving wherever and whenever possible.

Active and passive income

Another difference in income sources is the difference between active and passive income; the latter is more desirable than the former. Active income involves the exchange of hours for money; the most common form of active income is a wage paid for hours worked. As well as working yourself, you can increase the productivity of your time by expanding your farm and employing other people to work for you. You can also bring in active income by marketing and selling products, or by trading in financial instruments such as shares or foreign exchange on currency markets.

The problem with active income is that the capacity to increase your cash flow is limited by the number of hours available to work (there are, after all, only 24 h in a day), whether directly on your farm or in a business enterprise that you have to manage or supervise.

Passive income is generated by investments that don't require you to do much or any work. Examples include rent from real estate investments (residential and commercial), dividends from shares, and fees from licensing agreements.

Passive income is more desirable than active income because it is not limited by the number of hours available to you to work each day.

Off-farm employment and business income

It is common for one or more members of a farming family to have some off-farm earnings, either through work in a regional centre or through private contract work such as fencing, shearing, mustering or labouring. Via the internet, avenues for off-farm employment and revenue generation have increased dramatically over the last decade. For example, the internet facilitates direct sales of farm produce to consumers, especially for smallgoods such as specialist processed meats, eggs, herbs, honey and flowers, cutting out intermediate agencies that often pressure farmers into selling at lower prices than they would like. You can create your own website to sell produce in this way, using services such as Wordpress.com and Wix.com.

Off-farm investment income

Off-farm investments include commercial and domestic rental properties, shares, government bonds and fixed-term deposits. While real estate may ultimately provide a greater long-term profit on investment than shares, it also comes with a large up-front cost (generally a 20% deposit) and ongoing management responsibilities and maintenance costs. It is a long-term investment: with a 20% deposit, net rents and a little extra contribution should pay off a loan in 15 years. But, like government bonds, it is also not very liquid: if you need even some of the money, you will be obliged to sell the property to obtain it.

By contrast, investing in shares, which provide annual or biannual dividends, or in a dividend reinvestment plan involves minimal establishment costs apart from brokerage fees, and there are no ongoing costs or administrative responsibilities. Shares are also liquid, so you can add savings to your initial investment parcel at any time (in any amount over $500). Opening a share trading account is free and easy. However, as shown by the stock market drop in March 2020 in reaction to the COVID-19 pandemic, share value can vary quite dramatically – unless you are expert in playing the market, you would do best to invest in reputable (blue-chip) companies that pay regular dividends, and with a view to the long-term capital gain to be made from the later sale of the shares.

At the heart of an off-farm investment strategy is the goal of spreading risk by accruing assets that can increase your cash flow in times of financial stress on the farm, thus providing income that is not correlated with farm income. Keep in mind, however, that negatively geared real estate (i.e. property purchased with a bank loan) *increases* your overall financial risk profile. To reduce your financial risk, you must buy either shares or a property that can pay for itself from day one.

The livestock prices on my farm were very low during the millennium drought of 1997–2010. Happily, our low stocking rate meant that no animals died of malnutrition and I did not have to buy low-quality stockfeed at high prices. After infrastructure repairs, the profit and loss statement for that period was only a little above break-even point. Despite having a family to support, I was able to stay solvent using rental income from housing investments and dividends from shares. I had made all these investments between 1980 and 1997 – before the drought. During the drought, my balance sheet even expanded due to surplus income from off-farm investments, which enabled me to buy more rental properties, with resulting capital growth.

Had I used the profits that the farm had generated in better times to buy additional grazing land, I would have been much more exposed to debt during the millennium drought, especially if I'd bought land in the same, drought-prone region. Off-farm investment means sustainable farming in all its aspects.

Summary of key points

- To achieve long-term profitability as a farmer, you need to understand some key financial concepts, adhere to a predetermined budget, and be frugal rather than wasteful.
- Your farm budget should have five categories: (1) maintenance; (2) development; (3) debt reduction; (4) household expenditure; and (5) environment improvement.
- The combined totals for each category give you an aggregate cash-flow budget, which provides a summary of past and projected income and expenditure on a monthly basis.
- A cash-flow budget is essential to knowing well in advance when you might have a cash shortfall, which will allow you time to prepare for it. A cash-flow budget takes the stress out of money management and helps avoid costly impromptu decisions.
- Once you have finalised your budget, stress-test it by increasing your costs and decreasing your income by the same amount. What would you do if your financial situation became unsustainable?
- Loan repayment is usually the largest component of a farm budget, so pay off debt as soon as possible. Consider increasing your debt-reduction budget each month and plan to reduce debt as much as possible during profitable years. Avoid interest-only loans.
- Knowing your farm's annual rate of return means that you can compare your profitability from one year to the next and also compare your farm's profitability with that of other farms – including one you might be interested in buying – and other forms of investment.
- Increasing your bank borrowings reduces the equity you have in your farm. By projecting that debt into the future, you will see whether it will eventually improve your financial position or lead you down a debt tunnel to bankruptcy.
- Get into the habit of saving, no matter how small the amount.
- In years of surplus, gain a tax advantage by putting money into the federal government's Farm Management Deposits Scheme. This money can be redrawn in times of financial stress.
- Spread your risk by investing in non-farm assets that can be called on at times of financial stress on the farm.

Financial success provides peace of mind, which lets you to move from strength to strength in your farm business enterprise.

Endnotes

[1] A $100 000 loan at 7% interest compound over 15 years (factor of 2.759) = $275 900; $10 000 each year for 15 years compound interest at 7% (factor of 25.129) = $251 290, for a total of $527 190, rounded to $527 200.

2   Botsman R and Rogers R (2011) *What's Mine is Yours: The Rise in Collaborative Capitalism.* HarperCollins, New York.

3   https://hipages.com.au/article/how_much_does_rural_fencing_cost#How-much-does-rural-fencing-cost?

4   For example, direct seeding can be quick, cost-effective and low-risk compared to hand-planting tubestock. See www.greeningaustralia.org.au/wp-content/uploads/2017/11/FACT-SHEET_Direct-Seeding-benefits.pdf.

5   Lindenmayer DB, Michael D, Crane M, Florance D and Burns E (2018) *Restoring Farm Woodlands for Wildlife.* CSIRO Publishing, Melbourne.

6   For example, MYOB, Reckon and Xero.

7   Annual inflation in Australia is currently less than 2%; in the June 2019 quarter it was 1.6%. Source: https://tradingeconomics.com/australia/inflation-cpi.

8   https://au.investing.com/rates-bonds/australia-5-year-bond-yield.

3

Increasing your farm's profitability

The farmer is the only [person] in our economy who buys everything at retail, sells everything at wholesale and pays the freight both ways – J.F. Kennedy.

As with any business, your financial goal for the farm is to lower expenses and increase productivity and profit. Aside from the all-important budgeting (covered in Chapter 2), for optimal results you will need to do cost–benefit analyses of any projects that involve spending money. There is not always a linear and positive relationship between expenditure and the financial result. For some investments, benefits tend to increase to a certain optimal point as you increase your input, then reduce thereafter. This is called the law of diminishing returns (sometimes, the law of diminishing marginal returns) and is illustrated in Figure 3.1.

What Figure 3.1 illustrates is well known in economics: as inputs are added, output first goes up quickly, but with additional inputs the rate of increase slows and eventually output falls. This law of diminishing returns is clearly evident in farm production.

The application of fertiliser is an obvious example of the law of diminishing returns. If you apply too little fertiliser (the input), there will be no noticeable change to the pasture (output). However, applying more than you need for the job will be a waste of money because the paddock will show no extra improvement for the extra fertiliser applied. Even worse, you can harm your pasture through chemical burn caused by too much fertiliser. This principle can apply to many situations on a farm, as elsewhere.

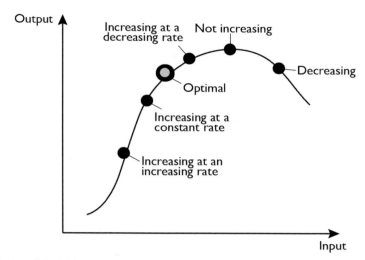

Figure 3.1: The law of diminishing returns

On the other side of a farmer's profitability ledger is the price you receive for the sale of your products. This is determined by the terms of trade of the commodity and thus is largely beyond your control. Nevertheless, sale prices have to be factored into your farm budget in the form of anticipated income from commodity sales.

Factors other than costs and benefits that affect farm profitability include levels of debt (which we discussed in Chapter 2) and the farmer's education level, age, business acumen and financial skills. Subjective qualities such as motivation and willingness to improve productivity are crucial to success. Factors such as annual rainfall are beyond your control, but it is not beyond your control to maximise your farm's capacity to capture and use water while minimising losses of it. The same goes for things like soil quality and health (physical, chemical and textural structure, evaporation, biodiversity and sunlight), and the ecological health and biodiversity of both pasture and non-pasture areas. The importance of environmental sustainability for your financial sustainability is discussed in detail in Chapter 5. Climatic factors such as drought and the frequency of adverse events are beyond your control, but being prepared for them is not (see Chapter 1).

Other factors affecting your farm's profitability include your access to labour when you need help on the farm (employees and family members/volunteers), and the distance from your farm to the market. First, I list some ways in which you can increase the efficiency of your farm. Second, I show you how to assess the financial effectiveness of such strategies, using specific examples. (To see how prepared you are for increasing profitability, you will find a useful checklist in Appendix 1B, 'Improving your farm's profitability'.)

Productivity improvements

Improving farm efficiency generally leads to productivity gains, whose value you can assess using cost–benefit analysis. Examples of changes that have resulted in productivity improvements on Australian farms over the last few decades include:

- the replacement of some types of labour with capital (i.e. with machinery) through mechanisation and automation (as well, the real cost of equipment has fallen thanks to tariff reductions), the use of simple, inexpensive tools and equipment to save labour, and a reduction of hired labour and higher farm-owner productivity
- reduced need for labour through the introduction of laneways for easier stock movement, cattle grids that allow faster movement around the farm, smaller paddocks that make stock mustering easier, double gates for facilitating stock movement, and improved wire spinners for easier and quicker fencing
- biodiversity improvements (e.g. the addition of tree plantings and riparian restoration) leading to more resilient pasture and shade for stock, which results in greater stock weight, reduced lamb mortality, cleaner water, and better ecosystem services
- structural improvements (e.g. the repair of cattle yards and fences) have meant easier and faster operations and fewer accidents
- paddock subdivision for better rotational grazing, leading to greater leaf area for photosynthesis in pasture
- tree plantings that provide shade, which leads to more fodder tonnage and thus more direct and indirect weight gains in stock
- greater water efficiency through more and better stock troughs that provide cleaner water and lead to greater weight gain, and reticulated water systems that enable more intensive use of available land
- improved genetics that have led to all studs producing more fertile bulls, and improved growth rates and food conversion in cattle (e.g. on my farm, weaning weights in 1973 were 220 kg and in 2017 they were 330 kg)
- pasture improvements and better fertilisers have led to less pasture damage and improved weight gain in stock
- improvements in cattle crushes and calf cradles, and better-designed cattle yards, mean labour savings, greater safety for workers and less injury to livestock
- reduced calving problems have resulted in less labour and fewer stock losses
- better electric fence technology has made it cheaper and easier to construct
- portable electric welders, generators and angle grinders, theodolites and cut-off machines (used for fencing, yard building and related farm structural engineering) save labour time

- downgrading the pipe and RHS steel supplies used in fencing and stockyard construction have made it more cost-effective
- bucket front-end loaders on tractors lead to efficiency gains, as do tipping trays on 4WD utes because they lessen labour costs.

Many farmers are thinking outside the square about ways to improve productivity and increase their bottom line.

Productivity gains through better water reticulation

Constructing a central water-distribution system using polythene pipes and troughs means that cattle don't have to walk far from a source of water to graze, which increases the stock-carrying capacity and productivity of a farm.

Increasing the stock-carrying capacity increases a farm's value and leads to greater capital gains.

However, each development case has to be evaluated according to cost–benefit analysis, or net present value analysis and the time value of money if there is a long timeframe involved. If you must borrow from the bank to complete the improvement, the interest payments might make it unviable.

Cost–benefit analyses

Let's say that a sales agent turns up with a ute-load of animal health products, including drenches to treat internal and external parasites in sheep and cattle. She has an array of glossy brochures about things like milking ability and weight gain, as measured by the company's empirical trials. The evidence all looks very convincing. However, every prospective project – even a relatively small one – must be subjected to a cost–benefit analysis. If you decided to take up all the new animal treatments, then you might gain little gross profit per head after all the costs are taken into account. If the timeframe for your project is too long and if you would have to borrow from the bank for the improvement, then the interest payments may well make it unviable.

A cost–benefit analysis of proposed items of expenditure helps you:

- compare potential benefits that would result from different expenditure options
- calculate what productivity gains have been made, which you can then compare with yearly terms of trade to see whether, overall, you are getting ahead.

Depending on the project you have in mind, spending that can be analysed might include the costs of labour, materials, stocking rate, supplementary feeding, fertiliser, drenching chemicals, price of land per hectare, pasture improvements and biodiversity improvements.

To do a cost–benefit analysis:

1. list the expected costs of each item
2. list the expected financial benefit from each item of expenditure
3. compare the gains using a baseline cost.

Calculating the expected financial benefits from extra expenditure gets easier with experience. If you are not sure about some things, talk to other farmers and do some online research about the benefits expected from certain products. If you do this every time you spend money, you'll soon see where your profit comes from and where you are wasting money.

The following is a simplified example of a cost–benefit analysis of sheep drench products. Say you have 2000 sheep to drench, and a product representative has told you that a new, more expensive, combination drench gets better results. But what exactly does 'better results' mean? Does it mean that you have to drench less frequently? Does it mean that more worms are killed at each drenching, with both short-term and long-term benefits because the treatment kills more worms now and also slows the development of drench resistance in the future?

If we focus only on the short-term benefits – lower stock loss from post-drenching worm outbreaks – then the cost–benefit calculation looks like this.

Say you have 2000 head of sheep and in previous years you've lost 0.5% of stock (10 head) post-drenching due to residual worms, which is a loss of $500 ($50 per head). Your calculations would be as follows:

- costs: the old drench cost $0.50 per head for a total cost of $1000, while the new drench costs $0.60 per head, for a total cost of $1200, meaning $200 in increased costs
- the estimated benefit which the product guide states has been scientifically demonstrated, is a 50% reduction in stock deaths: you lose five head of sheep instead of 10, × $50 = $250
- there is a $50 benefit to the new drenching product.

This is a reasonably simple assessment, but the calculation becomes more complicated the more aspects of the process you take into account, such as a reduced frequency in the need to drench and lower long-term parasite resistance.

Cost–benefit analysis allows you to assess different new products against each other. You can also use cost–benefit analysis in a more complex scenario, such as assessing what stocking rate is optimal for your land, based on the law of diminishing returns. If you want to apply a cost–benefit analysis to expenditure that is made over a significant period, then three more concepts must be included:

- the law of diminishing marginal returns
- the time value of money
- the net present value (NPV).

The law of diminishing marginal returns

As mentioned at the start of this chapter, the law of diminishing returns refers to the process by which the marginal benefit created through extra spending stops increasing and starts to decrease with each further investment. Put simply, after a certain point, the extra investment is both pointless and costly because it brings no extra gain.

In the case of fencing, for instance, you can work out how many workers produce the lowest labour cost per person per kilometre of fencing and what size crew is therefore the most financially efficient. Depending on the fencing team and the complexity of the job, a crew may be two to five people. The same goes for fruit picking: too many labourers get in each other's way and increase the load on employee administration; too few and the crop spoils before it is harvested.

When assessing the point at which returns on an investment begin to diminish on a particular project, you should first look at the equation under certainty. While seasons and markets can be very uncertain, you can start with expected values for various items, gathered from looking at results in previous years, talking to other farmers or doing online research. You should *always use conservative rather than optimistic estimates* to avoid projecting positive outcomes when in reality they might be neutral or even negative.

To go back to a fertiliser example: each extra sack of fertiliser has an additional price of, say, $10. One sack increases the wheat crop per hectare to five bushels, each worth $8, for a total of $40. That is, the fertiliser makes the land five times as productive as it was before (total output increases by a factor of five) and thus the extra income gained from the initial extra investment in fertiliser is $40. A second sack of fertiliser will increase the crop yield by an additional four bushels; this time, the extra income is $32 (4 × $8) in return for the extra cost of $10. When we get to the sixth extra sack of fertiliser, for which the marginal cost is $10, crop productivity increases by only 1.2 bushels for a total marginal gain of $9.60. Just before this is the point of optimum input and output, after which there is no increased return on the investment. The eighth level of fertiliser input will only increase output by half a bushel for additional total revenue product of $4, making it a waste of the money spent on the extra fertiliser.

As in all industry, two regular inputs into agriculture are labour and capital. Machinery costs have fallen in recent years, as have interest rates for loans to finance machinery purchases, while the cost of labour has increased. Therefore, some farmers have increased spending on capital equipment as a substitution for work that would previously have been done by a labourer. However, the use of machinery is also subject to the law of diminishing marginal returns regarding how much labour it saves. One unit of capital, say, a new fence-post driver mechanism, might save 10 days of labour but eight new pieces of machinery might save only an extra two days of labour. The marginal cost of new material of any kind should equal the cost of the labour reduction it results in. So, you need to calculate how much of your scarce cash to invest in new machinery.

Biodiversity returns on farms are also subject to the law of diminishing returns; for example, there may be an optimal number of tree plantings to be established on your farm. To assess the financial value of environmental programs specifically, economists talk about the marginal costs and benefits of abatement (the reduction of negative environmental issues such as pollution). The optimal level of abatement is where marginal cost equals marginal abatement, which means that any further increases in abatement will cost more than they are worth.

Say the optimal level of tree plots on your farm is 12 but you are short of capital, so you establish only four tree plantings. According to the principle of marginal benefits of abatement, you will get a proportionally high return on your investment in these plantings. If biodiversity projects on farms were subsidised, then the optimum level of environmental investments such as tree planting would be much higher. However, benefits accrued from this kind of work may not be seen for many years (at least five years and sometimes 10–20 years). In this instance, the time value of money must be taken into account in the analysis.

All the examples presented have involved the law of diminishing returns in conditions of certainty. But in agriculture, as you know, certainty is rare. The value of a crop or the sale price of livestock can vary considerably from year to year, and it is impossible to know what the weather will be like between planting and harvesting a crop. As a result, farmers must use the current prices of grain and livestock, together with average yields, to make a reasonable calculation on which to base their decisions. To a large extent, the decisions will be a gamble: they may be more or less right, but by the time you get your product to market, commodity prices might have shifted, sometimes significantly. As a farmer, you should make additional calculations, using best- and worst-case scenarios.

The next example analyses grazing and stocking rates. It demonstrates that increasing the stocking rate beyond an optimum point results in losses rather than increased profits.

Example 3.1: Cost–benefit analysis of increasing stocking rates

You should always avoid overgrazing because it leads to land degradation. Ideally, you should work out the optimum stocking rate for your land by determining where your marginal cost equals your marginal revenue at the point of maximum profit.

In the case of cattle and sheep grazing, the main inputs for your cost–benefit analysis will be labour, livestock, farm size, fertiliser, drench, supplementary feeding and biodiversity contributions. I don't include here the option of increasing the size of the farm because that will lead to an increase in debt, extra work and management. For this calculation, using figures from my own farm, we will stick with the given situation. Also, for simplicity, I have put biodiversity improvements into a separate basket. Thus, we have five factors to evaluate using the law of diminishing returns, all of which are related to the stocking rate using dry sheep equivalents (DSE) per hectare, the greatest of the five inputs.

If I keep my prime lamb flock at the same number and increase the number of cattle breeders by 20%, I would have 80 more cows and calves. Let's say that the extra capital required to buy this extra livestock is $200 000, for which the opportunity cost of not investing it elsewhere (e.g. in Treasury bonds) would be 5% or $10 000 p.a. The labour costs involved in keeping this bigger herd would increase by approximately $17 000, fertiliser for the extra improved pasture needed would rise by $8000, drench and animal husbandry by $8000 and supplementary feeding by $20 000. This adds up to an increase in total costs of $63 000 for the year – and that does not include the costs of extra management time or of greater damage to the farm environment. Increasing the number of cattle breeders by 20% would also make the farm much more vulnerable to the effects of drought and increase my personal stress levels.

If the calving rate remained at its usual average of 90%, then there would be 72 extra weaner calves, making a total of 432 head (up from 360). Weaning weights would fall by 20 kg because of the higher stocking rate: the original 360 calves at 310 kg gave a total of 111 600 kg @ $4/kg = $446 400, whereas the animals at the lower weight of 290 kg × 432 head = 125 280 kg × $4 = $501 120. This represents an increase in gross revenue of only $54 720 for the $63 000 in increased costs. In other words, the extra investment in stock numbers would cost me money instead of increasing my income.

Taking into account the increase in my management time and the possibility of environmental damage in addition to the fall in profit, the optimum stocking rate for my farm is 400 breeding cows and calves with replacement heifers. This is the point where marginal cost equals marginal revenue. This equation would change, of course, if there was a large increase or fall in the per kilogram live-weight of beef cattle or an increase or decrease in costs. A rise in prices for inputs such as stock feed would make the prospect of increasing stock numbers even less profitable.

Faced with falling livestock prices, many farmers try to increase stocking rates and therefore stock sales to maintain gross revenue. But a calculation of the law of diminishing returns shows that this is futile: as well as further reducing profits, it invariably causes more environmental damage.

Given the amount of paid labour on my farm, if my employees or contractors have spare time they can use it to establish plantings and repair stockyards, sheds and fencing.

At a meeting in Yass in February 2016, David Lindenmayer showed how a farm producing $100 of gross revenue for $95 in costs has a margin of $5, whereas a farm with revenue of $70 but only $55 in costs has a margin of $15. Sometimes, when we reduce output, there is an even bigger fall in costs and a bigger rise in net profit. We return to this question in Chapter 4, where we discuss the financing of farm expansion.

Time value of money and net present value

Cost–benefit analyses like those presented so far in this chapter are relatively straightforward when time is not included in the equation. However, as with large items of capital expenditure such as machinery, large buildings or long-term environmental improvement programs, the time value of money and net present value (NPV) should be included in such calculations. The buying power of a set amount of money, say $100, will usually decrease over any five-year period due to inflation (when inflation is at historically low levels as it is at time of writing, the decrease will be less than usual). If you invest $100 now, it will generate more money during the five-year period; if you don't have that money until five years into the future, then you have lost the opportunity to make that gain. This is called the time value of money.

In the case of $100, the value of the extra (marginal) money lost or gained is negligible, but if the base amount is $1 million then the value lost or gained is significant. For example, at just 1% interest, the gain on $1 million is $10 000, which grows to $50 000 over five years even without including the value of compounding interest. So, the time value of money should be included in any long-term cost–benefit analysis you do on capital expenditure projects.

The concept of NPV is used to compare the value of money now with the value of money at some point in the future. NPV is the present value of cash flows for your project at your required rate of return, compared with the value of your initial investment. NPV allows you to calculate your return on investment for any particular projected spending. By calculating the amount you expect to make from a project and translating those monetary returns into today's dollars, you can decide whether the project is worthwhile. This also enables you to compare one proposed project with another to see which would provide the greater profit.

To calculate NPV, you need to know your projected cash flows for each year and to set a realistic expected rate of return on the investment (this is shown formally in Appendix 3).

You can make these calculations using financial spreadsheets in Microsoft Excel; if you are not sure how to do this, there are tutorials on YouTube (www.youtube.com/watch?v=hG68UMupJzs&ab_channel=Excel%2CWordandPowerPoin tTutorialsfromHowtech) and other online sites on how to put your figures into the spreadsheet. Example 3.2 gives a detailed cost–benefit analysis with NPV to help with agricultural decision-making. In this example, a farmer wants to know whether it would be profitable in the short to medium term to sow elite pasture seeds on 100 ha.

Example 3.2: Investing in pasture improvement

Prudent investment typically requires a two-stage analysis. The first step is to determine whether revenues will at least cover the costs, and if so by how much. This is known popularly (although not accurately) as a cost–benefit analysis. If the investment is financed by a loan rather than available equity, then a further stage should be completed to check that revenues in each year will also cover at least the annual repayments on that loan.

Step 1: Example of a cost-benefit analysis

Investment projects can be presented using an investment timeline, as in Figure 3.2. The line begins at point zero (the present, now, today, etc.) because past costs and revenues are sunk costs, no longer relevant to an investment decision being made today. Any costs and revenues from now on are shown as upwards arrows (revenues) and downwards arrows (costs) for each year. A more detailed

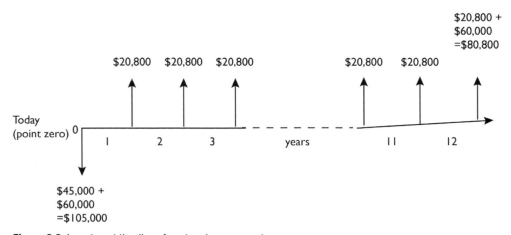

Figure 3.2: Investment timeline of pasture improvement

presentation could be broken into months or even weeks. Absence of an arrow indicates that no cost or revenue was incurred at that point in time. For simplicity, all cost and revenue arrows are shown to occur only at the end of each year, except for the initial capital investment which is shown as a cost at point zero.

A sheep farmer is considering intensifying the farm business and wants to know whether it would be financially beneficial to sow elite pasture species on 100 ha. The current stocking rate is six Dry Sheep Equivalents (DSE) per hectare. This particular pasture development would involve a once-only capital cost of $450 per hectare at point zero and have an expected lifespan of 12 years. The carrying capacity of the land would increase by an additional eight DSE per hectare, so a further once-only capital cost at the beginning would be the purchase of eight sheep per hectare at $75 per sheep: a total cost of $20 800 for the flock of 800 sheep purchased. To simplify the example, let's assume that no extra hired labour is required.

On the revenue side, the 'gross margin' per DSE is $26 p.a., where gross margin consists of income from the sale of wool and sheep, less variable costs (e.g. shearing, spraying, etc.), and excludes fixed or overhead costs such as depreciation, rates, interest payments or permanent labour. The gross margin for the flock of 800 sheep is shown as an upward arrow at the end of each year of $20 800 (800 × $26). All the 800 sheep are sold at the end of the 12 years for $75 each: this is shown at the end of the Figure 3.2 timeline as a combined upward arrow of a $60 000 sale (800 × $75), plus the annual gross margin for the 12th year of $20 800.

Note that amounts like the cost or sale price of a sheep of $75 are the same at the beginning of the 12-year period as at the end, and the annual gross margin of $26 per sheep also remains constant over the period. That is, we have assumed that inflation is at zero. It would be possible to forecast the actual cost of shearing or sale prices of a sheep in each of the 12 years, but only with difficulty, and significant estimation errors would be likely. That's why it is common practice to use constant cost and revenue values based on long-term historical averages. Equally, if the farmer is convinced that there will be a permanent increase in stock prices, then those higher values can be used.

Looking from point zero into the future, farmers would be aware that a gross margin of $20 800 received at the end of the first year is actually worth less to them than the same amount received today. The same sum received in the 12th year would be worth even less. A sum of $20 800 received at point zero can be spent immediately or placed in a bank to earn compound interest. If the farmer was able to earn, say, 6% p.a. from a friendly bank manager, then the $20 800 received today would grow into $24 773 ($20 800 × 1.06 × 1.06 × 1.06) at the end of the third year. To be happy with receiving money in the future rather than today, the farmer needs to be rewarded (compensated) with the larger amount. In other words, a dollar today is worth more than a dollar tomorrow.

Costs and revenues are incurred and received at differing times in the future. Because a sum received in the future is worth less than the same amount received in the present, all costs and revenues must be converted into so-called present values to make them comparable in today's terms. Reducing future values to present values can be done by applying the reverse of compound interest calculations; this process is appropriately termed 'discounting'.[1] Instead of multiplying the present value by the interest rate each year to obtain a future value, discounting involves dividing future values by the interest rate to convert them to a present value. So, an amount received in five years' time is divided by the interest rate five times. An amount received at the end of 12 years is divided by the interest rate 12 times, and so on.

Discounting future values in our example means that $20 800 received at the end of one year would be worth $19 623 ($20 800/1.06) today, and $20 800 received at the end of two years would be worth $18 512 ($20 800/1.06/1.06), and so on.[2] Table 3.1 converts the Figure 3.2 timeline to present values to illustrate this method. The interest rate used to discount future values should equal the highest feasible return that the farmer could achieve in other ways, as long as these other ways bear a similar level of risk as the pasture improvement project. An example might be the potential return from running a business sideline such as contract fencing or shearing.

By adding the present values of the gross margins achieved over the 12 years to the sale of $60 000 worth of sheep in year 12 ($191 154) in Table 3.1, the farmer has

Table 3.1. The present value of gross margins and the initial investment

Future value of gross margin ($)	Gross margin received at end of year	Interest rate used for discounting future values (%)	Number of times divided by interest rate	Present value of gross margin ($)
20 800	1	1.06	1	18 512
20 800	2	1.06	2	17 464
20 800	3	1.06	3	16 476
20 800	4	1.06	4	15 543
20 800	5	1.06	5	14 663
20 800	6	1.06	6	13 833
20 800	7	1.06	7	13 050
20 800	8	1.06	8	12 312
20 800	9	1.06	9	11 615
20 800	10	1.06	10	10 957
20 800	11	1.06	11	10 337
20 800	12	1.06	12	9752
+60 000	12	1.06	12	26 640
Sum of present value of all revenue received in years 1 to 12				191 154
Minus present value of initial capital investment				105 000
NPV				**86 154**

an estimate of the total revenue generated by the pasture improvement in today's terms. This amount can be compared directly with the value of the initial capital investment of $105 000 because it is spent at the same point in time (today). Because the investment is made at point zero, at the beginning of the first year, it is automatically expressed in present value terms and does not require discounting.

The difference between the sum of the present values of revenues and the present value of the capital investment ($86 154) is the NPV. A positive NPV, as shown in Table 3.1, generally indicates a worthwhile investment; however, you should further analyse the soundness of your estimates regarding the components of costs and revenues, as well as the potential effect of external factors such as drought and bushfires. Sophisticated statistical analysis such as Monte Carlo Analysis can be used to better gauge the probability of achieving a positive outcome when risk is taken into account.

Step 2: The financing analysis

The investment analysis described in Step 1 was used to check whether an investment like pasture improvement is inherently worthwhile, and it assumed that the initial investment of $105 000 is made using the farmer's own resources. However, the farmer may have to draw on savings or sell some livestock or old equipment in order to raise the amount required.

Now let's suppose the farmer must borrow the original $105 000 and is offered a farm development loan, to be repaid at 8% p.a. interest over six years. The yearly repayment to service the debt will be about $22 700. At this point, the cost of the debt repayments is higher than the annual gross profit earned ($20 800) from carrying the additional stock. In a low rainfall/low commodity price year, the repayments would indeed be hard to make. There is the additional risk that a quarter of the new pasture sown will fail. And there is penalty interest of 4% p.a. for late payment, which would worsen the situation.

To continue our hypothetical case: the farmer has a pre-existing bank loan on the farm of $500 000, borrowed over 15 years at 8% p.a. interest. There is no Farm Management Deposit or cash in the bank. The total debt grows to $605 000 with the addition of the extra loan for the new pasture project (an increase in total debt of 21%). The debt service on the original mortgage loan of $500 000 is about $58 400 in annual payments; when the repayments for the new loan are added, total annual repayments to the bank amount to about $79 200 p.a. ($58 400 + $20 800).[3]

It is clear that this farm intensification project is pushing the farmer towards a debt tunnel, resulting in more work and more stress. If the new development were financed with equity capital rather than debt capital, the outlook would be quite different. The equation would also change if the stocking rate were lifted by 10 DSE per hectare per year rather than by 8 DSE. All risk factors considered, this farmer would probably be better off not going ahead with the planned farm intensification and instead using the time to do off-farm work such as contract fencing. Alternative, low-cost improvements to the pasture might be spreading lime to reduce soil acidity and increasing the application of superphosphate which, if done gradually rather than in a one-off, large debt-financed capital development, would be more economically prudent.

The original pasture-improvement project would increase exposure to drought and lift the risk profile of the farm business. If the income generated by the farm were to fall significantly, funds available for servicing loans and paying for operating expenses could become insufficient. In other words, if the ratio of cash available to the farm in any particular year as a proportion of debt servicing obligations (interest and principal repayments) in that year – the debt service coverage ratio – were less than one, then the farmer would need to dip into savings or borrow more money to pay farm expenses.

Soil and pasture experts who sell chemicals and seed can present farmers with an array of impressive figures in support of their products, but farmers must protect their financial futures by doing cost–benefit analyses and quantified risk assessments, as well as monitoring their debt service coverage ratio. *The impact on debt of factors such as decreasing yield and increasing input prices are not always immediately clear, nor is the fact that the suggested extra profit can come with a large increase of risk of financial failure.*

Example 3.3 compares four mutually exclusive capital expenditure projects.

Example 3.3: Comparison of four alternative capital expenditure projects

A farmer wants to evaluate the following four mutually exclusive capital expenditure projects:

1. the establishment of tree plantings and restoration of waterways, including farm dam and stock shelter, costing $100 000 over 10 years ($10 000 p.a.)
2. spending $100 000 in Year 1 on updating equipment and implements
3. the immediate purchase of new, state-of-the-art cattle yards for $100 000
4. the immediate purchase of a new four-stand shearing shed for $100 000.

Option 1: Spending $100 000 on tree plantings and restoring waterways in 10 one-year instalments of $10 000 each

For this option, a discount rate of 5% applies, referring to the NPV and financial tables in business finance[4] ($10 000 × 7.7217 = $77 210). The time value reduces the cost. This environmental program, combined with a well-maintained physical structure, could increase the sale price of the farm in 10 years by 25% (an estimate of the differential between a well-maintained farm with environmental betterment and a rundown farm).

It may take 10 years for an environmental project to improve annual farm profitability. This must go into the NPV equation. A farmer's personal preference for environmental care should also be given a monetary value in the equation. Cost–benefit and NPV analytics also change greatly if there is a government subsidy for farm environmental restoration.

Option 2: Farm equipment update
The equipment will have a life of, say, 15 years after which it will only have trade-in value on the purchase of new machinery. There is also an operator safety issue, which should have a quantitative value applied to it. The productivity improvement must be quantified on an annual basis in the NPV equation.

Options 3 and 4: New cattle yards or shearing shed?
Both cases raise occupational health and safety issues for the farmer and hired staff, to which it is difficult to assign a financial value. You must calculate the productivity gains and increase in farm profit, as well as the increased market value of the farm in 10 years. Both these physical improvements will be subject to depreciation and become technologically outdated. All these factors need to be included in the NPV equation, along with animal injury mitigation, to find the best cost–benefit project. Animal injury is cruelty and an economic cost.

Conclusion
There is no right answer to the question of which option is the best, because the relevant figures will vary from farm to farm. Complex analyses such as these can be done with agricultural finance software, which can facilitate almost endless changes to the values of input and output parameters to help you to determine the option where the greatest productivity gains are likely to be made. Care must be taken to compare projects on an equivalent basis – as discussed above, projects over longer periods will accrue more annual benefits and costs. One method of making a valid comparison is to calculate equivalent annual values using the NPV results for each project (see https://en.wikipedia.org/wiki/Net_present_value).

Rural commodity price fluctuations
While we farmers can be disciplined and frugal with expenditure, it is very difficult to predict the price we will receive for our produce, whether it be live cattle or sheep, wool, meat or grain. Wildly varying commodity prices are largely beyond our control. Various price floor schemes have been tried over the years to reduce the impact of price fluctuations, but none of them has been successful. In very large commodity markets, attempts have been made to even out fluctuations in farm-gate purchase price by agencies artificially setting a price floor.

Historically, agricultural price stabilisation schemes have had varying degrees of success.

In Australia, the most notable failure of an attempt to set a floor price for an agricultural commodity was the Wool Reserve Price Scheme implemented by the federal government in 1971. The scheme accumulated a physical stockpile of wool bought from farmers during times of depressed prices, intending to sell it later in periods of high demand. The scheme was abandoned in 1991 after it was seen that the reserve price had been set too high and too large a stockpile was being amassed. When the system collapsed 'the price of wool fell overnight from 700 to 430 cents/kg, leaving the Australian industry with a stockpile of 4.6 million bales of wool (almost a year's production) and a debt of A\$2.7 billion'.[5] Some critics believe that the wool industry, whose production subsequently declined by 66%, has never really recovered from the debacle.[6]

The second major buffer stock scheme that attempted to stabilise commodity prices was the two-price arrangement for dairy products that operated from the 1970s to 1991. The problem was that the scheme set the domestic price too far above the export price, with the farmer receiving an average price. Like the floor price for wool, it was not particularly successful for the farmers. The same is true for other price-stabilisation schemes that have been tried in Australia.

Neither have farmers ever had any degree of success with the use of futures market contracts for beef, prime lamb and greasy wool. Forward delivery contracts for wheat have also had only patchy success. Indeed, some of the schemes have resulted in disastrous financial consequences for farmers.

The impact of monopsony power

Market concentration and market power apply when four or fewer firms control 80% or more of the market. Price-fixing and cartels are illegal under the Australian *Trade Practices Act 1974*, but this only applies to the sellers of products or commodities, not to purchasers. The two biggest supermarket chains in Australia – Coles and Woolworths – have purchasing power over the producer. Likewise, the 'Big Four' banks control some 80% of the lending market and thus can set the rates of interest and fees. They, like the two major supermarket chains, are price-makers.

By contrast, almost all 85 000 farm businesses in Australia are price-takers. With only a few mass purchasers of domestic rural commodities in Australia (compared with the numerous buyers of exported commodities) there is a reverse monopoly – a monopsony. Where there are four firms or fewer buying 80% or more of a commodity, they have immense purchasing power and can drive the farm-gate prices down because price-fixing by a purchaser is not illegal.

In the agricultural sector, Coles and Woolworths buy 80% of most farm produce. As well as the much-cited pressure put on the dairy industry by the '\$1 a litre milk' campaign by these two firms, there is industry concentration in the red-meat processing sector: only a few purchasers buy most beef cattle, mutton or lamb for slaughter. Thus, there can be deep gouging of farmers' incomes due to lack of competition in the post-farm-gate marketing chain.

Farmers cannot make production decisions quickly. Planting decisions about the acres to be harvested are made a year in advance; the same applies to the number of sheep, cattle and dairy cows that farmers stock. This means that the short-term supply is inelastic (i.e. supply does not and cannot change just because the price changes). In the longer term, a farmer can, in some cases, change the production mix between cattle, sheep or grain to make a more elastic supply. There can be a significant impact on farmers from even a small shift in consumer demand for farm commodities, because this will amplify price changes at the farm gate. When we add together the purchasers' market power, changes in final consumer demand and seasonal conditions in regional Australia, we get wide fluctuations in farm commodity prices.

Seasonal conditions such as low rainfall that reduce production of a given commodity, combined with a rise in overseas demand, can cause very steep price increases for that product. The reverse is true when there is a fall in overseas demand combined with good seasonal conditions that increase production of a commodity: the combination can trigger significant price reductions.

When there are only two or three companies selling the final farm product, if they increase the price to the end user, the consequence is a fall in the quantity of product sold. This fall in demand goes back to the farm gate in the form of a fall in the price paid. This use of their power is not only short-sighted of the managers of monopsony companies, it also raises questions about future national food security. The major post-gate problem for Australia's farmers is that there are just a few companies that make mass purchases of rural commodities. Lacking competition, these few buyers have monopsony power to continually drive down the price of rural products.[7]

The market power of a few companies is huge and must be addressed at a national political level. Together with highly variable and often low income for farmers, this situation created by all-powerful buyers is discouraging future generations from taking up farming. If the current arrangements continue, there will be many fewer farmers in the future. Those farmers who want to continue to farm must concentrate on what they can control more than commodity prices – that is, productivity.

Terms of trade and productivity gains

Farmers' terms of trade can be calculated as the ratio of the cost of prices paid for investments (e.g. farm machinery, casual labour hours) and the prices received for products sold (e.g. wool and livestock). Fluctuations in these prices lead to better or worse terms of trade, and ultimately to lower or higher profitability.

The costs of inputs are always on the rise because of inflation. International currency exchange rates also have a significant impact on costs. The prices of machinery and equipment in Australia are often strongly affected by fluctuations in the value of the Australian dollar against other major currencies, especially the US dollar, the euro and the pound sterling. When the Australian dollar goes down in value against these other currencies, buying equipment or machinery from other countries becomes more expensive.[8] Since at least the 1990s, most farm machinery in Australia has been imported, so the price fluctuations are beyond the control of farmers. The two most popular tractors – Massey Ferguson and John Deere – are produced by US companies. Between August 2011 and March 2020, the value of the Australian dollar against the US dollar almost halved, from $1.10 to $0.60, adding considerably to the cost of imported tractors.

Commodity prices for major export products such as wool and cotton, dairy, beef and veal, lamb, grains, oilseeds and sugar are also significantly affected by currency fluctuations, and by changes in global demand (as well as other factors such as competitive pressure from other producer countries like Brazil).

Within Australia, unless you are selling directly to a niche market, prices for your product are affected by the buying decisions of wholesale merchants and supermarket chains. They are also affected by climate-related fluctuations in yield: product shortages in drought years, for example, push up crop prices, and overall overproduction in the sector will push prices down in a good year.

Farmers have no influence whatsoever over currency values, but increases or decreases in terms of trade can be offset by productivity gains. Each Australian agricultural sector has experienced different price pressures. In the grazing industry, despite a considerable drop in terms of trade over the last 40 years, productivity gains have more than made up for the deficit.

Terms of trade calculations for my farm

To illustrate this general discussion of the terms of trade and their significance to farmers, I'll use the example of my own farm. I began working my property, Towong Hill, in 1973 and have kept records since then. The following discussion of my experiences with terms of trade is based on those records. My index of beef prices for the 45-year period from 1973 to 2018 changed from $0.65/kg in 1973 to $4.40/kg in 2018, a 5.23-fold increase in output prices.

For the purposes of this exercise, I have compiled an index of costs by combining the five main inputs essential to running the farm – wages, fuel, accounting fees, insurance, and vehicle prices. The base index of 100 in 1973 grew to 1344 in 2018, a 13.5-fold increase, or 6% compound per year. To work out the terms of trade, we must make the following calculation:

$$dx/dy = \text{change in input costs/over changing price of cattle}$$
$$= 13.44/5.23 = 2.57$$

What we see from this calculation is that there has been a 2.57 times deterioration in terms of trade since 1973. Put more plainly, *the real price received today for 1 kg of beef is only 40% of what it was in 1973.* The price of cattle increased more than five-fold in nominal terms (not real terms) and in the same 45-year period the terms of trade fell by 60%, representing a 2% p.a. compound depreciation.

So, for the grazing sector, terms of trade have deteriorated considerably since the 1970s. At the same time, however, the increased weaning weights of cattle, along with a reduction in labour and other input costs, has resulted in an estimated four-fold increase in productivity, especially that of top-performing farms.

Regarding this development, I agree with W.L. Ranken, a veteran stockbroker and corporate fundraiser, who said, 'The resilience of the farming industry since the Second World War in the light of declining farmer terms of trade is a big credit to the last three generations of farmers.' Many industries have had to produce more with less and increase productivity simply to remain competitive – and they have done so.

Calculations of terms of trade and productivity gains will vary from farm to farm depending on the commodities each farm produces. It is hard to quantify productivity improvements because they, too, vary from farm to farm, but in the livestock industry you can multiply the weight-gain factor by the stocking-rate factor by the labour-productivity factor and then divide that sum by fall in the quantity of material inputs to arrive at the total productivity increase. On my Towong Hill farm, there has been an approximate 3.6-fold productivity increase in 45 years, representing a 3% p.a. compound improvement.

Summary of key points

- To make productivity gains and improve your chance of making a profit, you must decrease your costs and increase the value of your products.
- To know whether you are making productivity gains, you must do a cost–benefit analysis for each of your items of expenditure.
- To do a cost–benefit analysis you need to: (1) list the expected costs of each item; (2) list the expected benefits gained from expenditure on each item in financial terms; and (3) compare the gains against a baseline cost.

- More sophisticated cost–benefit analyses take the law of diminishing marginal returns into account. This is the point at which the marginal benefits gained from further investment in an action stop increasing and start to decrease per item of investment.

- A good example of the law of diminishing marginal returns is the application of fertiliser: if you add too little fertiliser there may be no noticeable benefit to the pasture; add too much and the additional money you've spent on fertiliser is wasted as the paddock shows little extra improvement beyond the optimal point of application.

- Time should be included in cost–benefit analyses of large items of capital expenditure such as expensive machinery, large buildings, pasture improvement, or long-term environmental improvement programs.

- The time value of money theory recognises that a sum of money is worth less in the future than it is worth today because of inflation. You can use an NPV formula to calculate your expected return on an investment in a particular project over a period of time. This allows you to compare the potential benefits of one proposed project with those of another (e.g. building new infrastructure versus pasture improvement).

- While farmers can be frugal, prices paid for their products are generally out of their control and vary considerably from year to year. Past collectivist attempts (e.g. setting a price floor on markets) to even out farm-gate prices have largely failed in Australia.

- The major problem for Australia's farmers is that there are simply not enough mass-purchasing companies of rural commodities. Without competition, the dominant supermarket chains have the power to drive down farm-gate and sale-yard prices of rural products.

- While overall terms of trade have been reduced for Australian farmers over the last few decades, productivity gains may have made up for most or all of this deficit.

Many other factors can affect a farm's profitability, such as the farmer's education level, age, motivation and business acumen, level of farm debt, distance to market, and farm-specific geographical factors (e.g. soil health and rainfall). Chapter 4 addresses some of these factors.

Endnotes

1 The interest rate used for discounting is often called the discount rate.
2 The process of discounting a long series of future values by hand can be tedious. A shortcut is to discount each of the previous results once by the discount rate. For example, the present value of $20 800 received at the end of the second year needs to be divided only once by using the Year 1 Figure ($19 623/1.06), which yields $18 512. The present value of $20 800 received in Year 3 would then be equal to $17 464 ($18 512/1.06). An alternative is

to multiply by the discount factors in tables available from websites such as https://www.cimaglobal.com/Documents/Student%20docs/2010%20syllabus%20docs/P1/P1-performance-operations-tables-2010-syllabus.pdf. Spreadsheets in Excel can automatically calculate the sum of the present values in all years.

3   Annual loan repayments can be calculated using annuity tables to estimate an equal amount for each year of the loan period. Annuity calculations are explained at https://smartasset.com/retirement/annuity-table. In the table shown, the annuity factor for a 15-year loan charging 8% p.a. is 8.559. Dividing the loan amount (its present value when initially invested) of $500 000 by 8.559 yields $58 418, which is the required annual repayment of interest and principal.

4   Peirson G and Bird R (1976) *Business Finance*. McGraw-Hill, Sydney, p. 435.

5   McRobert K (21 February 2019) *Briefing: Dairy regulation and floor pricing*. Australian Farm Institute. https://www.farminstitute.org.au/briefing-dairy-regulation-and-floor-pricing/

6   Murphy D (30 July 2011) Shrunken industry fleeced by politics and greed. *Sydney Morning Herald*. www.smh.com.au/entertainment/books/shrunken-industry-fleeced-by-politics-and-greed-20110729-1i4at.html.

7   *ABARES Insights*, Issue 1, 2020.

8   For example, if the Australian dollar is worth US$0.69, then a farmer will need to spend A$1.46 for every US dollar of equipment (e.g. a US$50 000 piece of machinery will cost A$72 760).

4

Financing farm expansion

Buying neighbouring land: double your profit but double your risk?

Large farms are defined by the Australian Bureau of Agricultural and Resource Economics and Sciences (ABARES) as those with receipts above $1 million per year in real terms. The number of large farms increased from 3% to 15% of the total number of farms between 1979 and 2019. In the same period, the output of the large farms grew from ~25% to 58% of total national farm production. Successful farmers have expanded their holdings, buying up land from unprofitable or retiring farmers. This can lead to increased productivity and gains in the environmental care of farms. But buying more land also can be risky because instead of diversifying income sources, farmers can be exposed to co-location problems, commodity price variations and the pressure of servicing overcommitted bank loans.

In many rural communities, the social pecking order is based on the number of acres you own. This can add to the temptation to expand your holdings. However, buying a neighbour's farm can be a financially and environmentally high-risk strategy as the income/loss is strongly correlated with the home farm. This makes it emotionally stressful for underfinanced farmers in times of economic hardship in the region or the industry. You should consider very carefully whether expansion is a good strategy for improving your financial returns.

If the average return on capital of a farm is 1.5% p.a., then you would do better to diversify by investing in commercial property or shares, which currently have an average income yield of 4%.[1] A higher yield is possible if you do your research and make a careful selection of stocks. This is 2.7 times more profitable than a farm and far less work by any measure. In some rural communities, stock market investing is frowned upon. However, if your mean income yield is 1.5% and you borrow at 7% to extend the farm the financial problem is clear. The situation also leads to more work and more stress.

In the case of grazing farms, buying a neighbour's property will not necessarily lead to gains from economies of scale by having two farms in the same region. What might work for cropping, with mega-paddocks and super-sized machinery, will not necessarily work for grazing businesses. (While most of the examples in this book are drawn from grazing experience in northern Victoria, the principles are generally applicable to any farming operation.) The reasons are many, but typically include travel time for the farmer and workers, and transport of stock and machinery between properties. In addition, a large farm – one that cannot be run by just one family – whose stockyards and shearing sheds are long distances away and may include a second or third set of yards, requires employee labour with wages and on-costs, all of which can severely lower profitability. As many farmers know, a sure way to cause yourself grief is to employ permanent labour on your own account. Today, most farmers use only casual labourers when they are needed.

More significant, however, is the risk presented by local climatic events, which can massively increase the likelihood that you will lose everything you have rather than just part of your agricultural enterprise. You may lose your livelihood. Many farmers have lost all their farming operations to a single major weather event. In the catastrophic south-eastern Australia fires of December 2019–January 2020, some farmers around Corryong who had bought up neighbouring farms lost everything on two or three properties. Banana growers in far north Queensland lost vast plantations to Cyclone Larry in 2006. Five years later, having not diversified geographically since the original devastation, they lost the lot again to Cyclone Yasi in 2011.[2]

So, even if you think you can make economies of scale through buying up a neighbouring property, you need to balance that against the greatly increased risk of adverse events wiping out your whole business.

Is the property worth buying?

When a neighbouring farm comes up for sale, your first step is to assess whether the property is worth buying from a financial standpoint. Leave aside any emotional notions about the desirability of the place and ignore what anyone else says about it being a good or bad investment proposition. Do your sums and talk to your accountant!

In Example 4.1, I set out the calculations you should make before deciding whether to buy more land in a given set of circumstances. You can extrapolate some of the calculations and considerations to your own situation if you are thinking about buying more land.

Example 4.1

- The farm is 200 ha and the purchase price is $1 million, including stamp duty.
- The farmer requires an 8% return after capital gains tax. I am not including an owner-operator salary in this equation, and for simplicity, have excluded income tax from the calculations. However, you will need to include capital gains tax at the top marginal rate.
- The farmer is 45 years old and plans to retire at 60 (i.e. a 15-year planning period).
- The property needs some improvement, so a development budget of $120 000 over four years for infrastructure and environmental investment is included in the calculations.
- For simplicity, the cost of further livestock purchases is not included at this stage, but a second-stage analysis would include it.
- The expected future value of the land is $2.5 million (sale price) with approximate selling costs of $100 000. A low sale price would be $2 125 000 and a high one would be $2 875 000. This is calculated on the basis of a standard deviation of 15% ($375 000) either side of the mean; the probability of any future sale price falling within this range is high.
- The farmer will make no profit from the extra land for four years, then will make $25 000 for each of the next 11 years. One standard deviation of profit of 50% ($12 500) is possible; it is a large variance, but typical of farming.
- Net present value is calculated at an expected $2 500 000 and an expected profit of $25 000 per annum from Year 5 onwards.

These calculations could also be subjected to the high and low values, to stress-test the investment returns. In a second-stage analysis, the possibility of drought and the resulting reduction in profit would have to be taken into account. All these variations can be put into a computerised spreadsheet in Microsoft Excel or some free or commercial agricultural financial software.

The calculation tells us that in this case, there is a near-perfect positive correlation with the profit levels of the farmer's original farm and thus economies of scale can be achieved with the addition of the neighbouring property, with the work done using existing equipment and labour.

The annual return on the investment after four years is the value of the property plus the value of capital improvements, divided by the estimated annual income:

$$\$1\,120\,000/\$25\,000 = 2.23\%.$$

As a comparison, the current grossed-up yield for Commonwealth Bank share dividends is 8.61%, which is nearly four times the projected return on the farm. Dividends have much less variance than farm profit and company dividends have no correlation with farm profit unless there is a big shift in the macro-economy.

However, the farmer needs to make three further calculations, for (1) capital gains tax, (2) net present value and (3) where the money will be borrowed, the amortised loan debt service.[3] The results of the three calculations reveal that not only is buying this additional land under these financial conditions not a good idea, but also that even if no debt were incurred to make the purchase, it would still be financially risky. If a 50% loan were required, then the purchase becomes unthinkable. Unless the farm is in excellent condition and requires no capital improvements, there is no reason to believe that buying it will improve the farmer's overall profitability, especially if the current farm is also in debt. The purchase would create more of the same: the same asset and therefore the same risk correlation with problems of drought, fire, disease, etc. but at double the scale. For those who are interested, the details of these calculations are laid out in Appendix 3.

Further financial considerations

Doubling the size of the property you own might double your output and revenue with little extra equipment and labour, resulting in increased profit. But any farmer thinking of this kind of expansion must compile a careful profit and loss statement for the additional land, based on expected seasonal rainfall and commodity prices. A purchase of land for $2 million with a loan of $1.2 million over 15 years at 6% interest (making a debt service ratio of 10.13) will mean annual loan repayments of $121 500.

The revenue earned from greater sheep, wool and wheat sales might equal $250 000. Additional fixed and variable running costs for the new property might be $100 000, reducing the profit to $150 000. Return on capital with full equity (i.e. with no mortgage) would be $2 000 000/$150 000, that is, 7.5%, which is certainly on a par with returns from commercial property investments. However, while this is a positive example of increased economies of scale, even with no extra mortgage, there is little cash left after the loan repayments (and $28 500 of family expenses must be deducted).

In this example, the farmer now has a new large mortgage, which increases the risk profile of the entire farm business and increases the volatility of household income. All the farm's financial eggs are in one (larger) basket, without any commodity or regional diversification that would spread the financial risk at times of drought and low commodity prices. Both the farmer and the family have more work and more stress.

The following is another example of the impact of increased debt, incurred to buy additional land, on the volatility of farm household income.

Let's say that the expected profit of Farm 1 with no debt is $150 000, although in a drought year that expected profit falls to $50 000. With the purchase of another farm, the expected combined profit of the two farms would be $300 000 minus debt repayments of $121 000, leaving roughly $179 000 of expected household income.

This looks good, but in a drought year the expected profit of both farms is only $100 000, less the debt repayments of $120 000, which results in a $20 000 loss. This is an unacceptable variation in household income and one that dramatically increases the risk profile of the farming business.

Affording loan repayments

Even if your bank manager or loans officer – with their eyes on bonuses, lending targets and bank profits – approves your application for a loan to buy additional land, this does not mean that you can afford the repayments. The loans officer knows that you have good equity in the initial farm and that the bank can charge penalty interest and significant foreclosure fees if the new farm fails.

Table 4.1 shows how an unserviced debt will grow in 10 years if it incurs penalty rates of 10% interest. It clearly shows how a loan that goes unpaid will chew up farm equity very quickly. You may gain greater social approval when you have a bigger farm, but if you are paying 5% interest on the loan and the return on your investment in the larger farm is only 3%, then you are in negative territory and you could end up losing both of your farms. Add to this the volatility of

Table 4.1. Debt after 10 years of 10% interest on loans from $200 000 to $2 million

Original principal ($)	Debt after 10 years ($)
200 000	518 740
400 000	1 037 480
600 000	1 556 220
800 000	2 074 960
1 000 000	2 593 700
1 200 000	3 112 440
1 400 000	3 631 180
1 600 000	4 149 920
1 800 000	4 668 600
2 000 000	5 187 400

annual cash-flows due to variation in rainfall and poor commodity prices, and you and your family will soon be on a rollercoaster ride, with the danger zone starting at around $500 000 debt.

In some ways, the agricultural entrepreneurs who are willing to take on the risks involved in expansion are the future of farming. A few good years of rainfall and high commodity prices might yield a profit of, say, 12% on a big wheat or sheep farm and thus quickly pay for the purchase of additional land. Some farmers enjoy the positive feeling of building up land holdings.

But for most, buying additional land is a high-risk gamble against drought and low commodity prices. If you're planning to borrow to expand your farm, then the best strategy is to buy in small parcels, paying off a small acreage before you borrow to buy yet more. Prepare a very careful budget and have your accountant check your figures. Be prepared for a rough, though potentially rewarding, ride if you can pull it off. Above all, make sure that your own farm is paid off first and, if possible, have off-farm investments in stocks or real estate. Make sure to stress-test your income statement using lower commodity prices and the costs of a drought, but also stress-test your loan calculations as shown in Table 4.2 at 10% or 12% interest, not the current 5% or less. Interest rates are currently at record lows and so may well rise.

Commodity prices have not kept up with inflation, whereas land values have risen at a higher rate than inflation. As a result, farm profitability as a percentage of land value is lower today than it was 70 years ago. However, rising farmland values over the long term mean that farmers who expand their holdings may build a very large net worth over a 40-year period. Therefore, keen farmers in their 20s will find it a rewarding challenge.

Borrowing on farms is a risky business because interest rates are up to 3% higher than on a mortgage for an urban owner-occupied house, to compensate the bank for the higher risk involved.

Table 4.2. Amortised loan repayments (principal and interest) over 20 years

Debt interest	3.5%	6%	8%	10%	12%
Debt service ratio	7.04%	8.60%	10.04%	11.59%	13.22%
Total debt ($)	Annual repayment ($)				
250 000	17 600	21 500	25 100	28 975	33 050
500 000	35 200	43 000	50 200	57 950	66 100
750 000	52 800	64 500	75 300	86 925	99 150
1 000 000	70 400	86 000	100 400	115 900	132 200
1 500 000	105 600	129 000	150 600	173 850	198 300
2 000 000	140 800	172 000	200 800	231 800	264 400

Annual compound interest at various interest rates and annual repayments (rounded).

Let's say that a farm runs a cash deficit (including living expenses) of $80 000 p.a.; a debt of $600 000 that goes unserviced and thus incurs penalty interest rates of 10%, compounded annually, will grow into $1 556 220. Add this to $80 000 cash deficit each year:

$$\$80\ 000 \times 15.9374^* = \$1\ 274\ 992$$

* The multiplier for compound interest.

Add these together and in 10 years the $600 000 debt will have grown to $2 831 212 – *almost five times the original sum*. The power of compound interest is working against the farmer and rapidly chewing up farm equity.

The next equation is time and the rate of compound interest. Does the compound rate work for or against a farmer? Obviously, the longer the mortgage period at a higher rate of interest, the bigger the debt in the end. A $200 000 debt at 10% interest that goes unpaid for 10 years will amount to $518 740. Over 20 years, again at 10%, that will be $1 345 499. These calculations assume compounding annually, not monthly.[4]

Calculating your farm business equity

As I advised in Chapter 2, your annual review of farm operations and finances should include a calculation of your farm business equity. To do this, first add up your total assets:

Total assets = land + plant + livestock + water licence + off-farm investments + cash at bank + Farm Management Deposits

Then add up your total debt:

Total debt = mortgage + credit card debt + outstanding invoices + vehicle lease costs

Divide the debt by the assets and multiply by 100 to get a percentage debt-to-equity ratio. For example, assets of $1 250 000 and debt of $150 000 is a 12% debt ratio, whereas assets of $1 250 000 and debt of $850 000 is a 68% debt ratio. *With a debt percentage of greater than 25% or a farm equity of 75% or less, you are in the debt-tunnel danger zone.*

Many farmers, particularly young ones, overestimate income and underestimate the expenses of owning a farm. This is why you should have your figures checked by an accountant who is experienced in rural matters before you buy a farm or you take on an expansion.

Tax minimisation and negative gearing

Many farmers try all possible ways to reduce their gross profit to reduce their tax. They often do this by purchasing expensive equipment (often upgrading a perfectly good machine such as a tractor or combine harvester to a new model) or buying more land, rather than paying off debt with post-tax dollars. This strategy can backfire spectacularly – the expenditure on new machinery doesn't reduce a farmer's financial risk, it increases it. When a bad drought or other adverse event occurs, the high level of debt can result, either rapidly or slowly but surely, in the bank foreclosing on the farm.

For example, $500 000 profit spent on a new combine harvester, which immediately loses up to 30% of its value immediately after purchase, will not necessarily contribute significantly to increasing the farm's productivity. It certainly doesn't contribute directly to paying off debt.

If, instead, the farmer pays tax on the $500 000 profit, the proportion remaining (e.g. $275 000 at the full marginal tax rate of 45%) could be used to purchase some or all of an off-farm investment such as real estate, shares or government bonds. With the original investment growing annually through rental income, dividends or bond interest, the farmer will not only eventually make up for the original tax paid, but also have a source of income for ever.

It is also good to remember that paying tax is our way of contributing to the society in which we all live and from which we all benefit directly through roads, municipal services, hospitals and education for our children. There is no good reason why farmers should not claim legitimate deductions and depreciations, but farmers spending profits simply to avoid paying tax is not only financially imprudent, it also verges on immoral.

Negative gearing of an investment property, whether it be residential or commercial real estate, also increases a farmer's financial risk profile by adding more debt. Either the investment should be purchased with sufficient equity that it is cash-positive from the start, or the available money should be invested in the stock or bonds markets until sufficient capital has been accumulated to make a cash-positive real-estate investment possible.

Optimising your farming operations

If you are still uncertain about whether to increase your farming operations by buying more land or adding large-scale capital infrastructure, the following general principles may help. Even if you are not contemplating expansion, the points made here are good to keep in mind when you are compiling your annual budget and farm plan review. You should also refer to the checklist 'Expanding your farming operations' in Appendix 1D.

The optimal size of your operation

For any type of agricultural enterprise – grazing, cropping, dairy, orchards, intensive horticulture or niche markets – you can work out the optimal size of the operation using cost–benefit analysis, the law of diminishing returns and economies/diseconomies of scale. This optimum may be based on:

- the stocking rate, i.e. the number of animals per hectare
- in the case of grazing farms, the amount of land that a family can operate on their own
- in the case of poultry and pigs, the number of sheds
- in the case of feedlots, the number of yards
- in the case of a dairy herd, the operational capacity of the dairy farm.

Losses will be incurred at anything larger than the optimal size, and there will be no good economies of scale for anything smaller. Of course, size will also be dictated by rainfall and inherent carrying capacity.

You need to determine the optimal size for your type of operation in your district. This will involve talking to neighbours, your own accountant and other accountants in the region. Most accountants have a good idea of how various farms in a district are doing financially. You can also do more general research by studying data provided online by agencies such as the Australian Farmers' Federation and ABARES.

The optimal way of managing your operation

One of the major constraints on the size of a farming operation is the amount of labour needed to run it efficiently and effectively. Once a farm increases beyond the capacity of the farming family to run it, myriad problems arise associated with finding and employing other people to work on the property.

In the case of a very large farm or in an industry in which seasonal labour is essential, farm hands are necessary. But a family farm that doubles in size through the purchase of additional land might end up with more problems than the expansion is worth if most of the profit goes on paying workers and half the principal farmer's time is spent on finding, engaging, supervising and doing payroll for that larger labour force.

The degree of mechanisation is also a consideration in the way farms operate. Expensive machinery may cost more to buy or lease than the benefits it brings in reducing labour requirements. You should always do a cost–benefit analysis before buying expensive machinery.

The optimal way of financing expansion

As I have stressed throughout this book, the best way to finance an expansion to your operation is to buy it outright, with cash you have in the bank, supplemented by off-farm investments that will stabilise income and act as a reserve should unexpected events demand extra money.

The optimal level of motivation

Doubling your responsibilities by taking on another farm of similar or equal size to the one you're already managing, or embarking on a major capital infrastructure project such as a new poultry, milking or piggery shed, can add considerably to your personal stress levels.

An optimal level of pressure can provide motivation without crippling you with anxiety about how the logistics will work out, or whether your venture will ultimately be profitable. An optimal degree of pressure is of course a highly personal matter, but everyone has a limit.

Meticulously calculating the financial impact of your planned additional investment and having your calculations checked by an accountant before making a decision will help address your doubts. You must be able to sleep at night and to

Summary of key points

- Expanding farming operations by buying a neighbouring property is a goal for many farmers but this strategy is financially dangerous because it involves no diversification of income source. Both farms will suffer the same commodity price variations and, if there is a catastrophic event, are likely to suffer loss of infrastructure and income.
- Economies of scale might be achieved with more hectares of land and better use of labour and equipment, but the neighbouring farm may require significant investment in stocking, infrastructure, weed control and pasture improvement.
- It could take four or more years to start turning a profit on unfamiliar land, and you may need to borrow more money. Bank loan repayments start from Day 1.
- To determine whether additional land is worth buying, you will need to calculate the capital gains tax, net present value and, if you need to borrow money, the amortised loan debt service.
- If you are borrowing money to buy additional land, you must calculate whether the loan repayments will be affordable in a variety of circumstances, including higher than expected interest rates and lower than expected income.
- The return on capital for small to medium farms is usually lower than it is for off-farm investments such as shares or real estate. Research alternative

investments to see which will be the best for you financially, in both time spent and the amount of stress incurred.

- Minimising tax by purchasing unnecessary, expensive equipment or through negative gearing a property, is a poor financial strategy. Profits should first go towards paying off debt, then put into savings or small investment options such as bonds or shares, then into larger investments such as residential or commercial real estate. You can optimise your farm operations by working out the optimal:
  - ➤ farm size for your industry sector
  - ➤ way of managing your operation
  - ➤ way of financing an expansion to your operation
  - ➤ amount of personal pressure that motivates you without overburdening you with stress.

face your partner in the morning with a clear head and continuing enthusiasm for what you've taken on.

The bottom line is that to make your farming business profitable, you need to have an optimum size farm, operating at optimum carrying capacity, with no debt and minimised expenses, while putting as much of your savings as possible into off-farm investments.

Endnotes

[1] Mickleboro J (2020) 4 high yield ASX dividend shares for income investors. *The Motley Fool*. www.fool.com.au/2020/01/03/4-high-yield-asx-dividend-shares-for-income-investors/.

[2] Rural Industries R&D Corporation Industry Overview, *Focus on Cyclone Resilience Research and Development*. Publication No. 13/122. www.agrifutures.com.au/wp-content/uploads/publications/13-122.pdf.

[3] A similar example can be found in Chisholm AH and Dillon JL (1966) *Discounting and Other Interest Rate Procedures in Farm Management*. University of New England Press, Armidale, pp. 15–18.

[4] Financial formulas taken from Peirson G, Brown R, Easton S, Howard P and Pinder S (2012) *Business Finance*. McGraw Hill, Sydney, Chapters 3–4. Author's own calculations.

5

Achieving environmental sustainability

Healthier farms make for healthier farmers (www.sustainablefarms.org.
au/about).

Improving farm profits through environmental investment

When farmers aim purely for short-term profit, environmental degradation can
lead the farm to the point where productivity drops, even if you keep your stocking
rates at the same (high) level. Such a decline will be especially rapid at the onset of a
drought and is likely to lead to the need to completely destock, resulting in a
significant loss of income.

Even minimal overgrazing of farms or poor management of cropping lands
will eventually lead to degraded and eroded soil, weed infestations, dryland
salinity, poor water quality and the siltation and eutrophication of downstream
areas of waterways. Your challenge as a grazier is to determine the optimal
stocking rate at which profits are maximised and the land remains productive in a
long-term, environmentally sustainable way.

Many farmers are sceptical about the benefits of lowering stocking rates, spending
money on environmental remediation and fencing off areas of native vegetation. In
2003, the situation was summed up in this way by the Australian Farm Institute:

*The continuing tension between the productive objectives of farmers and the
environmental objectives of governments arises because of the [inevitability]*

of these two sets of objectives coming into conflict. While maintaining a small proportion of farm area for environmental purposes may not greatly reduce farm output, eventually as the proportion of the area required for conservation increases, productive output decreases. Because of this inevitable tension, the more governments regulate to achieve environmental outcomes on private land, the more perverse incentives are created for farmers to minimise environmental features of the land.[1]

In some areas where state governments have introduced 'command-and-control'-style conservation legislation, land values dropped as much as 20%. Another perverse outcome was the incentive it created for farmers to hide or even eradicate threatened species on their land.

Many farmers resent government bodies' implicit expectation that farmers should bear all the costs of protecting the environment, which sometimes leads to losses of productivity for farmers while accruing benefits for the broader community. Even the obligation to put as little as 2.5% of land aside for conservation can have a cost impact on a farm business.[2]

However, as well as setting aside biodiversity conservation areas, good environmental stewardship of rural land involves managing all biological, soil and water resources in a sustainable manner. The initial aim should be to remediate any degraded environmental assets, and then enhance them.

Some innovative and courageous farmers have shown that conservation and productivity gains can be achieved simultaneously through innovative farming practices.[3] In addition to improving the resilience and quality of pasture, regenerative agriculture techniques can reduce the need for pesticides, herbicides and fertilisers, which both reduces costs and further improves environmental sustainability. Enhanced biodiversity can also lead to direct improvements in ecosystem services such as bee pollination of crops and control of agricultural insect pests.

Finally, when farmers make obvious environmental improvements such as fencing riparian zones, planting healthy resilient pasture and fencing tree plantings, they can later sell their land for up to 25% more than a neglected farm would bring.[4] In itself, this gain offsets the infrastructure costs of environment protection and biodiversity enhancement.[5]

Proven strategies for achieving a more profitable and environmentally sustainable farm include:

- being open-minded about adopting new farming practices
- being flexible about change until you work out what's best for your farm
- researching and attending workshops on farm sustainability
- becoming an active member of industry-specific farm sustainability project groups (e.g. regenerative grazing groups)

- becoming an active member of a local Landcare group
- cooperating with neighbours on environmental works projects
- accessing government support programs such as the NSW Farm Innovation Fund
- completing regular, scientifically based monitoring and testing, and keeping records of the results to assess progress
- adopting regenerative grazing principles, with holistic grazing management and an increase in perennial grasses in grazing pasture
- regularly assessing pasture resources and destocking early at the apparent onset of drought
- protecting and rejuvenating riparian and wetland areas though stock exclusion, bank rehabilitation, removal of willow trees etc.
- trialling new approaches to natural asset management and adopting cutting-edge agricultural technologies to get the best results
- actively managing feral pest animals, especially rabbits
- actively managing wooded pastures, and restoring and managing biodiverse woodlands
- reducing financial risk and increase profits by diversifying on-farm income streams and selling direct to market through specialist cooperatives.

You can find further information on sustainable farming by reading about demonstration farms such as those documented on the ANU Sustainable Farms website.

Cost–benefit analyses of environmental improvements

The benefits of environmental improvements on farms are many and varied. However, from a financial point of view, you should complete a cost–benefit analysis of any proposed expenditure to be sure you'll be getting the best value for your money and some idea of the future returns on your investment.

The following example shows a cost–benefit analysis for a 45-year-old farmer who intends for his farm to be in top order, with minimal noxious weeds, optimally managed waterways, trees and environmental and physical farm infrastructure. The farmer will have to address waterway erosion, plant trees and be up-to-date with the latest knowledge about improving soil; for example, by applying gypsum to reduce acidification and improve soil physical structure.

To quantify this improvement project, we will assume that the farm will be sold in 20 years' time rather than be willed to the next generation. The eventual sale of his farm for a 25% premium above the sale price of a rundown one will provide the farmer with extra funds for retirement and the possibility of extra money to help his children establish themselves in other occupations.

The cost–benefit analysis shows that if the farm sold for a $500 000 premium in 20 years' time, this would equate to a 12% compound return over the period, which is an excellent financial result. The analysis does not include any increase in production over the 20-year period, which would also add value to the investment.

While location and the quality of soil play a critical role in farm management value, anecdotal evidence suggests that an estimated premium of 25% for a farm that is in good shape and has been environmentally improved is not out of the question. The standard of the house on the farm is important because a house in poor condition may discourage potential buyers; however, the increasing numbers of corporate/investor/overseas buyers are likely to place more importance on environmental factors, soil quality and infrastructure standard than on the quality of the house.

Intergenerational environmental improvement and financial health

All of us – individuals, governments and their agencies, corporate bodies – must make decisions about what we will consume now and what we will leave for future generations. Using renewable resources at a faster rate than they can be reproduced means leaving less for future generations, as does using non-renewable resources and not recycling what we use. Consuming now in a way that leaves less for the future is, I believe, ethically wrong.

In an agricultural context, responsible choices about the use and protection of resources include leaving our farms in a good state for future generations, whether they are inherited by our own children or not. The tasks involved in maintaining and improving farm environments include tending to the physical and chemical quality of the soil, improving the quality of pasture, controlling erosion and reducing noxious weeds and feral animal populations. Profligate spending on machinery or additional land instead of on the continual improvement of the quality of your farm's natural resources, uses up the rightful inheritance of future generations.

There is a close relationship between being financially responsible and environmentally responsible: someone who buys a car through a loan at 10% interest ultimately pays more for the car than someone who delays their purchase until they can pay in full. The same goes for a farmer who wants to increase productivity quickly and ends up overgrazing or overcropping; the consequent poor soil will ultimately reduce productivity – and thus the overall value – of the farm (it increases the farmer's discount rate). The farmer who increases productivity and wealth by sacrificing current consumption or improving the environmental and other health of the farm will end up with a more viable and valuable farm.

Environmental improvements and mental health

Farming is an inherently risky business and the financial performance of a farm is often strongly linked to the farmer's emotional health. When you become a farmer, you are gambling with both human and financial capital in an uncertain world, facing variation in income, losses from changing seasonal conditions, fluctuating commodity markets and terms of trade deterioration. The stress generated by years of drought and low commodity prices takes a toll on family relationships. Farmers who experience severe depression are less able to run their farms effectively, and the whole family suffers.

Anxiety, depression, trauma, anger and even shame felt by farmers and their families as a result of financial stress are often amplified by social isolation and by a culture that reveres stoic resilience, particularly in men, at any cost. Recent efforts have been made by government and community organisations to change this culture, but it will be a long, slow process.

Some farmers handle uncertainty better than others, but anyone subjected to long-term financial stress will inevitably suffer deteriorating mental health. Yet, the stigma of failure and discrimination against mental illness are worse in rural and regional areas than in the city.[6] Farmers have an unacceptably high suicide rate – and that of male farmers and farm managers is about double the rate of the general working population.[7] Failure in farming can also be fatal for other health-related reasons,[8] and some farmers can end up as defeated, dysfunctional people.

However, farmers and their families can take some preventive and ameliorative steps to manage and alleviate depression in a rural environment. Strategies for reducing stress include selling burdensome assets, clearing up the farm and planting some trees.

Sell burdensome assets

If you have incurred large debts, then the obvious thing to do is to sell some of your land, tough as the idea seems. While it may result in the loss of some economies of scale, it will reduce your stress and that of your family: you will have less debt and less land to look after, which will likely make your agriculture more sustainable and be better for the environment than owning extended assets that you can't manage properly.

You might fear a reduction in your standing in the community if you sell land. Ignore unsupportive people and seek out those who will stand by you. After calculating the pros and cons, including your capacity to maintain debt, have a realistic talk with your accountant and make decisions according to what will help you to become both more financially stable and mentally strong and resilient.

Clean up your farm

Small things about life on the farm can be uplifting. If you can afford not to sell some land, then you can perhaps also take the less drastic but highly productive step of repairing farm sheds and fences with second-hand materials and ensuring that all gates swing free. Clean up the inside of your sheds, sell unused tools and equipment, clear away rubbish and sell scrap metal to a recycling business. Seeing the results of this kind of work can give you an immediate sense of achievement and motivate you to do more.

A bank manager or loans officer can generally tell if a farm is credit-worthy just by coming up the front drive. Weeds, rubbish and collapsing sheds signal a lending risk long before the lender look at the income statement and balance sheet. A well-kept farm, where waterways have been restored and trees planted, will impress the bank manager. It is also uplifting for you and your family who, after all, live every day on the property.

Plant some trees

Planting trees can be uplifting. In my experience, trees and native animal life are good for both the soul and the bank account – and they also make me feel proud when I'm showing guests around. In my opinion, we farmers are only life custodians of the land: we need to make the most of the time we're here but also to try to leave it a better place than it was when we acquired it.

Trees look good, increase the value of the farm, improve bird, mammal, reptile and insect life and, very importantly, provide shade for stock. The health of livestock, particularly black cattle and sheep burdened with wool, improves considerably when they have greater access to shade. Healthy livestock equals more weight, and more weight equals greater profit per head.

Search out direct subsidies for preparing paddocks for tree plantings and fencing materials and hiring of equipment at reasonable rates. Look into what help Landcare Australia and Greening Australia can provide.

Get your family to help you to establish tree plantings – it's good exercise and good for family cohesion.

Develop an interest in the benefits of biodiversity. There is an abundance of online information about tree and shrub planting as a way of investing in the improvement of farming environments. Several groups, such as the Environmental Farmers Network, provide examples of innovative biodiversity improvements. As shown by the Sustainable Farms project at ANU, a small sum of money can result in a top return on investment of your time, in the form of increased vegetation, tree cover, greener pastures and bird calls. These can improve the mental health of farmers, our sense of wellbeing and our pride in what we do.

Summary of key points

- If you manage a farm purely for profit without a thought for the future, then it will eventually deteriorate environmentally to the point where productivity is reduced.
- Your challenge is to determine the optimal stocking rate for your farm, which maximises profits while the land remains productive on a long-term, environmentally sustainable basis.
- Investing in the environmental quality and biodiversity of your farm can lead to improved profits as well as a more attractive farm, better psychological wellbeing and social good.
- Farms with a high level of environmental sustainability and biodiversity are much more tolerant of and resilient to drought.
- Natural regenerative pasture techniques reduce the need for pesticides, herbicides and fertilisers, and thus result in cost savings. Enhanced biodiversity can lead to direct improvement in ecosystem services.
- A well-maintained property with obvious environmental improvements can sell for up to 25% more than a neglected farm.
- There are many proven strategies for achieving an environmentally sustainable farm that also increase the farm's profits.
- Do cost–benefit analyses of any proposed expenditure on environmental works to ensure you'll get the best value for money and thus have some idea of your future financial return from the investment.
- It is unethical to leave a farm in a worse state than when you took it over. Profligate spending, instead of investing in the natural resources of your farm, affects the next generation's natural inheritance.
- The farm balance sheet and your emotional health are often strongly interconnected. Anxiety, depression, trauma and anger created by financial stress can be amplified by social isolation and a rural culture that still reveres stoic resilience, particularly among men.
- Stress has detrimental, sometimes fatal, effects on farmers' health. There is an unacceptably high suicide rate among farmers. If you have burdensome debt, it might be time to think about divesting yourself of it. Cleaning up your farm, protecting remnant vegetation and planting trees will bring benefits to both your farm and your mental health.

Endnotes

1   Australian Farm Institute (2003) *Farming Profits and the Environment*. Occasional Paper, June, p. 1. www.farminstitute.org.au/publications/occasional-papers.html.
2   Miles CA, Lockwood M, Walpole S and Buckley E (1998) Report No. 107, Charles Sturt University, Albury. Cited in Australian Farm Institute (2003) *Farming Profits and the*

Environment, Occasional Paper, June, p. 2. http://www.farminstitute.org.au/publications/occasional-papers.html.

3   Lindenmayer DB, Michael D, Crane M, Florance D and Burns E (2018) *Restoring Farm Woodlands for Wildlife*. CSIRO Publishing, Melbourne.

4   A study reported in *The Age* (Saturday 27 February 2016) by David Parnell of the University of Western Australia states that developed mature vegetation can add 25% to the value of a farm.

5   Walpole S, Lockwood M and Miles CA (1998) *Influence of Remnant Native Vegetation on Property Sale Price*. Report No. 106. Johnstone Centre, Charles Sturt University, Albury.

6   See, for example, www.abc.net.au/news/rural/2015-10-09/country-hour-mental-as-panel-discussion/6841036 and www.abc.net.au/news/rural/2017-10-12/suicide-still-a-bigger-problem-in-the-bush/8996568.

7   www.abc.net.au/news/2008-08-19/farmers-suicide-rates-double-national-average-study/482170.

8   Death rates among male farmers and farm managers aged 25–74 (for the period 1999–2002) were 33% higher than those of the general male population the same age. Lockie S (2015) *Australia's Agricultural Future: The Social and Political Context*. Report to the Australian Council of Learned Academies, p. 21. https://acola.org.au/wp/PDF/SAF07/social%20and%20political%20context.pdf.

Conclusion

Farming is one of the most difficult occupations in Australia. Few other jobs require such resilience, constant innovation and adaptation to ever-changing conditions – the environment, markets and commodity prices. Yet agricultural production is a fundamentally important activity – it is crucial to Australia's food security and to ensuring that the country can make a contribution to feeding humanity globally.

However, we are all too aware that the financial challenges of farming can be considerable. Good financial decisions are therefore crucial, as we have seen in so many ways in this short book, whether they are about buying additional fertiliser, changing the drenching chemical being used, purchasing a neighbour's property or dividing the family assets as part of a succession plan (including deciding whether to keep or sell the farm that has been in the family for years).

Some of the decisions that farmers face are likely to become even more weighty in the future, for a range of reasons. Demand for food will rise due to an ever-increasing human population, as will demands for that food to be produced in ecologically sustainable and animal welfare-appropriate ways. It is inevitable that climate change will bring higher temperatures, more droughts and more fires, and that commodity prices will remain highly volatile. Land prices are increasing rapidly, with knock-on effects for levels of farm debt. The age profile of most farmers is also rising, highlighting the significance of succession planning and of intergenerational equity among the farmer's children. The current record low interest rates cannot be expected to be permanent; they must surely rise eventually by several percentage points. These (and many other) concerns all point to the pressing need for farmers to be financially literate. This is why we are presenting this book: to help farmers become more aware of the financial implications of their actions, to better assess the costs and benefits of alternative approaches to farm management, and to understand better whether the loan they are contemplating is affordable and can be justified in terms of its planned use.

If this book can help even just a handful of farmers to manage their properties better, to reduce the levels of stress associated with their lives and to help 'save the family farm', then we will consider that we have made a worthwhile contribution to agricultural Australia.

John Mitchell
(with David Lindenmayer and Bruce Chapman)
January 2022

Appendix 1: Your farm finance checklists

A farmer's footsteps are the best fertiliser on a farm. Routinely check vehicles, machinery, sheds, stock, feed, yards and fences.

I have devised the following seven checklists to help farmers make financial decisions that improve the management of their farms, leading to increased productivity and, ultimately, to increased profitability of their farms. You can copy and use them in your own financial planning and management processes.

The checklists apply in the following contexts.

Appendix 1A: Buying your first farm
Appendix 1B: Improving your farm's profitability
Appendix 1C: Shock-proofing your farm
Appendix 1D: Expanding your farming operations
Appendix 1E: Creating a farm business succession plan
Appendix 1F: Improving your farm's environmental sustainability
Appendix 1G: Working with government.

Appendix 1A: Buying your first farm

This checklist is helpful if you have never owned a farm before and are thinking of buying one. Tick each item as Yes or No. You should seriously consider the issues raised in any items you mark No. *If you say No to 10 or more items on the list, then you may not be adequately prepared to buy the farm you are thinking about.*

Prior knowledge and experience	Yes	No
I come from a farming family.		
I have some practical experience in farming.		
I have experience in farming in this property's industry sector.		
I've run a business before.		
I've employed people before.		
I'm familiar with farming conditions in this district.		
I know how to use accounting software.		
Level of debt	**Yes**	**No**
I can buy this farm outright without taking out a bank loan.		
I have saved enough money to need a loan of less than 25% of the purchase price.		
I have enough money to purchase necessary stock and equipment.		
I have enough off-farm income to support my family until the farm becomes self-sustaining.		
I've saved enough money to support my family until the farm becomes self-sustaining.		
I won't need an overdraft during the first year to keep the farm solvent.		
Planning	**Yes**	**No**
I understand the need for a comprehensive farm plan, including a business plan, an environment plan, adverse-event plans and a succession plan.		
I've already drafted a basic farm plan for this purchase.		
Budgeting, financial management and business evaluation	**Yes**	**No**
I keep my personal and business accounts up to date and submit my annual tax returns on time.		
I have sufficient financial knowledge to manage the farm accounts competently.		
I know how to use farm finance software or can afford an accountant or bookkeeper to use it.		
I've prepared a budget for running the farm for the first five years.		
My budget includes money for maintenance, development, debt reduction, household expenditure and environmental remediation.		
I've stress-tested my budget by increasing costs by 5% and decreasing revenue by 5%, then continuing to increase costs and decrease revenue by that amount until I get to the point of insolvency. As a result, I know what latitude I have for budget shortfalls.		

Continued

Appendix 1A. *Continued*

I've considered the financial advantages and disadvantages of buying this farm.		
I've asked for and received the owner's balance sheets for the farm for the last 5 years.		
The owner's balance sheets are in line with my estimates of the annual cash flow generated by the business.		
My accountant has confirmed that the owner's balance sheets are a reasonable record of the farm's financial transactions.		
My accountant/I have audited the condition and value of the farm's infrastructure.		
My valuer/I have audited the condition and value of the stock/crops included in the farm purchase terms.		
My accountant has reviewed the farm accounts and approves of my decision to buy the farm.		
Productivity and profitability	**Yes**	**No**
My accountant/I have calculated what the rate of return on capital is for the farm.		
I know how to carry out a cost–benefit analysis.		
I know I must carry out cost–benefit analyses on all major items of expenditure if I buy this farm.		
I understand the concept of the time value of money.		
I have a good idea of how long it will take before the farm is profitable.		
I understand the general principle of how productivity can be improved and have some ideas about how I can improve productivity on this farm.		
I have a good understanding of how, when and why changes in commodity prices occur in this sector.		
I have a good understanding of how, when and why changes in input costs occur and what impact they will have on the profitability of the farm.		
Farm management	**Yes**	**No**
I know how much labour I will need at various times of the year.		
I understand the government's rules and regulations for employing people.		
I understand the cost in time and money of employing people.		
I've studied how to optimise the management of this farm and how to run it more efficiently.		
I believe I can manage this farm differently to improve its productivity.		
Adverse events	**Yes**	**No**
I understand the consequences if this farm is partially or destroyed by a severe adverse event such as fire, flood or cyclone.		
I know the probability of drought in this region.		
I have enough money to fully insure the farm for all adverse events.		
Environmental remediation	**Yes**	**No**
I've consulted an expert in the government or private sector about the environmental condition of the farm.		
I understand the environmental condition of the farm and what needs to be done to either maintain that standard or improve it.		

The farm is in good environmental condition.		
Some environmental work, like the planting of tree plots or riparian fencing, has already been done on the farm.		
I know what the likely costs will be for carrying out necessary and supplementary environmental work.		
I know what ecosystem services are.		
I understand the likely environmental damage that will result from overstocking this farm.		
Government	**Yes**	**No**
I understand what government legislation applies to this piece of land (e.g. the *Environment Protection and Biodiversity Conservation Act 1999*).		
I know what development regulations apply to this farm.		
I know what environmental regulations apply to this farm.		
I know what industry-specific regulations apply to this farm.		
I understand what animal welfare regulations apply in this industry sector.		
Mental health and relationships	**Yes**	**No**
My mental health is strong.		
I do not suffer from chronic depression.		
I've compared the personal advantages and disadvantages of buying this farm and believe that the advantages outweigh the disadvantages.		
I acknowledge that the possible pressures involved in buying this farm may increase my stress levels.		
I acknowledge that running this property may impinge on my time with my partner and family.		
I have the full support of my family.		
I know an experienced mentor who has agreed to guide me.		
I'm already a member of a relevant industry group.		
I want to integrate into and contribute to the local community.		
I am not buying this property for increased status.		
I'm not being pressured by someone else to purchase this property.		

Appendix 1B: Improving your farm's profitability

This checklist is for you if you want to improve your farm's profitability. You need to pay greater attention to the items that you mark No.

Planning	Yes	No
I understand the importance of setting goals and measuring the success of outcomes – profitable or otherwise.		
I have set out project goals for the next five years in a farm business plan that includes labour and material needs, their estimated cost, and likely dates for proposed expenditure.		
I have worked out a farm environment plan that includes cost–benefit analyses for major project items.		
I have developed adverse-events plans and understand how an adverse event would impact my capacity to become or remain profitable.		
I have a succession plan that is as fair as possible to all relevant family members and states clearly what is to happen to the farm in the event of my unexpected death. A copy of this plan is with a lawyer/accountant or other trusted person and I have appointed someone I trust as my executor.		
I have had clear discussions with all relevant family members about what will happen in the event of my death or retirement from the farm.		
Budgeting and financial management	**Yes**	**No**
I have enough financial knowledge to manage the farm accounts competently.		
I know how to use farm finance software or can afford an accountant or bookkeeper to use it.		
I keep my personal and business accounts up to date and submit my annual tax returns on time.		
I have a comprehensive farm budget that includes allowances for maintenance, development, debt reduction, household expenditure, environmental remediation and an aggregate monthly cash-flow budget.		
I update the six components of my budget on a monthly basis.		
My aggregate budget shows when I should expect a cash shortfall so that I can prepare for it.		
My farm has been operating at a loss, but I have a good idea of when it will be profitable again.		
I know the importance of being frugal and sticking to budgets.		
My focus is on decreasing my costs and increasing the value of my farm outputs.		
Level of debt	**Yes**	**No**
My level of debt is manageable; the farm income covers loan repayments and other business and family expenditure.		
I have a manageable level of debt and am not slowly going backwards.		
My level of debt doesn't hamper my ability to make productivity gains.		
I know my annual rate of return on capital and whether my farm's annual profitability is increasing or decreasing.		

Cost–benefit analyses	Yes	No
I know how to do a cost–benefit analysis and use the law of diminishing returns to work out the optimum point of investment to get the best results for the money I spend.		
I regularly complete cost–benefit analyses on all major items of expenditure.		
Either my accountant or I do a cost–benefit analysis to evaluate competing options for major items of expenditure.		
I understand the meaning of the time value of money and my accountant/I can apply a net present value formula to calculate the expected return on my long-term investments.		
Productivity improvements	**Yes**	**No**
I understand the general principles of how to improve productivity.		
I have some ideas about how I can improve productivity on my farm.		
I've looked at optimal ways to manage my farm more efficiently.		
I have a good understanding of how, when and why changes in commodity prices occur in my farm industry sector.		
I've prioritised options for managing my farm to improve productivity.		
I understand that many factors contribute to productivity improvements, including my own knowledge, experience and level of education, my business acumen and my mental health and motivation.		
Environmental remediation	Yes	No
I know that environmental degradation on my farm contributes to loss of productivity.		
I know that improving the farm environment could lead to improved productivity.		
I've estimated the likely costs for completing necessary and supplementary environmental works and have factored them into my budget.		
I have a list of ways in which I can bring in extra income that will pay for environmental remediation work.		

Appendix 1C: Shock-proofing your farm

This checklist is designed to help you to arm yourself against direct financial shock from rising costs and loss of income, and from adverse-event shock. Adverse-event shock will result in financial shock unless you are adequately prepared with insurance, savings and off-farm investments.

Financial shock	Yes	No
I've assessed the risk of financial shock from such events as crop failure, stock disease, commodity market decline, loss of off-farm income, drought or a rapid-onset adverse event.		
I have a realistic budget that I follow strictly and update monthly.		
I update my assets and liabilities inventories quarterly.		
I do not purchase any major item for the farm without carrying out a cost–benefit analysis to ensure it will result in improved productivity.		
I'm making regular repayments on my debt as scheduled.		
Neither I nor other family members are profligate with money.		
I reduce my debt whenever I can by making extra repayments on my loan in times of surplus.		
I know what level of income I need to bring in and how great a decrease in income will leave me with cash-flow problems.		
I have a good professional relationship with my accountant and am not afraid to seek advice on the financial status of my farming operations.		
I have a good professional relationship with my lending institution and am prepared to let my loans officer or bank manager know early if I'm experiencing financial difficulties.		
Adverse event: drought	Yes	No
I have a comprehensive drought plan and feel adequately prepared.		
I regularly do things to increase the capacity of my farm to deal with drought.		
I monitor rainfall regularly and keep rainfall data for my farm.		
I monitor water resources on my farm on a quarterly if not monthly basis.		
I monitor medium- to long-term weather conditions on a weekly or monthly basis so I have an early warning that the farm may encounter drought conditions.		
I have drought 'insurance' in the form of a large reservoir of stockfeed.		
I'm aware of the implications of the changing climate in my region.		
I'm prepared to sell stock early to preserve pasture.		
Adverse event: fire, flood and cyclone	Yes	No
I've researched the likelihood of an adverse event impacting on my farm and have read survival stories of other farmers to understand the magnitude and impact of what could happen.		
I have comprehensive adverse event plans for fire, flood and cyclone (as appropriate) and feel adequately prepared.		
I've taken action to reduce the risk of a catastrophic event destroying infrastructure, stock and crops on my farm.		

	Yes	No
I regularly monitor short-term weather conditions, so I have an early warning of an adverse event.		
I have a family survival plan, including early evacuation of family, pets and farm service animals.		
I belong to an industry association or local cooperative that assists with adverse-event planning and business recovery.		
Insurance	**Yes**	**No**
I have comprehensive farm insurance that covers loss of income and a place to live in the case of an adverse event, as well as loss of stock and infrastructure.		
My farm insurance covers me for fire damage.		
I'm in a flood zone and my farm insurance covers me for flood damage.		
I'm in a cyclone zone and my farm insurance covers me for cyclone damage.		
Savings and off-farm income	**Yes**	**No**
I have sufficient money put aside in long-term deposits or in the Farm Management Deposits Scheme to sustain me and my family for long enough to get the farm back to financial profitability after a severe adverse event.		
I have sufficient bank savings to sustain me and my family for long enough to get the farm back to financial profitability after a severe adverse event.		
I have enough capital reserves in long-term investments in shares or real estate that produce income to sustain me and my family for long enough to get the farm back to financial profitability after a severe adverse event.		
I have sufficient off-farm income to enable me and my family to survive a drought indefinitely.		

Appendix 1D: Expanding your farming operations

This checklist is useful if you own a farm and are thinking of buying another property. Any items marked No in this list should be taken seriously. If you answer No 10 or more times you should question whether you are adequately prepared for the additional responsibility involved in buying more land.

Risk assessment	Yes	No
I know that buying another farm in the same district increases my exposure to adverse events such as drought, fire, flood and cyclone.		
I know that buying another farm in the same industry sector increases my exposure to adverse commodity fluctuations.		
I know that buying another farm in the same industry sector increases my exposure to industry-specific shocks such as disease or industrial sabotage.		
I know I might not achieve the economies of scale I've calculated by increasing the size of my holdings.		
I know that if the farm business I buy fails for financial reasons, I might also risk my current holdings.		
I've sought the opinion of an independent evaluator who has confirmed that there are no legal or financial elements in this property purchase that would present a risk of later complications.		
My accountant has provided a full cost–benefit analysis of the farm purchase over the life of the anticipated loan period, and I understand the level of financial risk I will be taking on.		
My accountant has provided a comparison of the likely returns from investing in this farm versus investing in an off-farm investment such as shares or real estate.		
Level of debt and financing the new investment	**Yes**	**No**
I have enough money to buy this new farm outright without a bank loan.		
I have enough money to require a loan of less than 25% of the purchase price.		
I've calculated that I'll be able to comfortably afford the loan repayments.		
I have enough money to purchase necessary stock and equipment for the farm.		
The new farm will not be a financial drain on my current farming business.		
I will not need an overdraft during the first year to keep the combined farming operations solvent.		
Planning	**Yes**	**No**
I already have a comprehensive farm plan – including a business plan, an environment plan, adverse event plans and a succession plan – for my current farm holdings.		
I've drafted a basic farm plan for the new enterprise.		
Budgeting, financial management and business evaluation	**Yes**	**No**
I keep my business accounts up to date and submit my annual tax returns on time.		
I have enough financial knowledge to manage the new farm accounts competently.		
I know how to use farm finance software or can afford an accountant or bookkeeper to use it.		
I've prepared a budget for the first five years of the new venture.		

My budget includes allowances for maintenance, development, debt reduction, household expenditure and environmental remediation.		
I've stress-tested my budget by increasing costs by 5% and decreasing revenue by 5%, then continuing to increase costs and decrease revenue by that amount until I get to the point of insolvency. As a result, I know what latitude I have for budget shortfalls.		
I've compared the financial advantages and disadvantages of buying this farm.		
The owner has given me balance sheets for the last five years of the farm's operation.		
My accountant/I have audited the condition and value of the farm's infrastructure.		
A stock valuer/I have audited the condition and value of the farm's stock and/or crops that are included in the sale.		
I've evaluated the optimal size of a farm for this industry sector. My current holdings combined with this farm (if they're in the same sector) fit within this optimum.		
My accountant has reviewed the farm accounts and approves of the purchase.		
Productivity and profitability	Yes	No
My accountant/I have calculated the rate of return on capital employed for this purchase.		
I regularly perform cost–benefit analyses regarding aspects of my current farm operations.		
I understand the concept of the time value of money and the timeframe required to achieve an adequate return on this new investment.		
My accountant has confirmed my calculations for long it will take for this farm to become profitable.		
I know how farm productivity can be improved and have some ideas about how to improve productivity on this farm.		
I have a good understanding of how, when and why changes in commodity prices occur in this farm's industry sector.		
I understand that the need to focus on this new farm might interfere with my ability to maintain productivity on my current farm.		
Farm management	Yes	No
I've calculated what labour I will need at various times of the year for this new farm.		
I already have experience in employing people.		
I've looked at how I can manage this farm more efficiently.		
The way I intend to manage this farm will result in productivity improvements.		
I've assessed the optimal way in which this farm should be managed and there are things that can be changed to achieve this optimum.		
Adverse events	Yes	No
I understand the consequences of this farm being destroyed by fire, flood or cyclone.		
I'm aware of the probability of drought in this region of Australia.		
I have sufficient money to fully insure the farm for any adverse events likely to affect it.		

Continued

Appendix 1D. *Continued*

Environmental remediation	Yes	No
This farm is in good environmental condition and I've assessed what I need to do to maintain that standard.		
This farm is in poor environmental condition and I have assessed how to bring it up to the standard necessary to improve productivity.		
Environmental improvements, such as the planting of tree plots, has already been done on this farm but more needs to be done.		
I've calculated the likely costs for carrying out both essential and supplementary environmental works on this farm.		
Government	**Yes**	**No**
I understand what government legislation applies to this piece of land (e.g. the *Environment Protection and Biodiversity Conservation Act 1999*).		
I know what development regulations apply to this piece of land.		
I know what environmental regulations apply to this piece of land.		
I know what industry-specific regulations apply to this piece of land.		
I know what animal welfare regulations apply to this industry sector.		
Mental health and relationships	**Yes**	**No**
My mental health is strong.		
I do not suffer from depression.		
I've assessed the personal advantages and disadvantages of buying this additional property and believe that the advantages outweigh the disadvantages.		
I acknowledge that the potential additional pressures of managing additional, unfamiliar land may increase my stress levels.		
I acknowledge that the additional pressure and time involved in managing this additional, unfamiliar land may impinge on my time with my partner and family.		
I have the full support of my family in this new venture.		
I'm not buying this additional property to increase my status in the community.		
I am not being pressured by someone else to purchase this additional property.		

Appendix 1E: Creating a farm business succession plan

No matter what age you are, it is important to have a farm succession plan and legal documents such as a will, contracts, deeds and enduring power of attorney to support that plan. If you're young and don't intend to sell your farm in the foreseeable future, you still need to document exactly what will happen should you die unexpectedly or become incapacitated in such a way you can no longer manage your farm. All your succession documents should be clearly marked and gathered in a readily accessible place (off the farm) that a trusted friend or family member knows about.

Family discussion	Yes	No
I've had detailed discussions with my family about what will happen to the farm business if I die suddenly or am incapacitated.		
My family agrees about what will happen to the farm business if I die suddenly or am incapacitated.		
I'm planning to retire in the next five years and have had detailed discussions with my family about the best way to divest myself of the farm business.		
Key members of my family are in agreement with my succession plans.		
I understand the possible negative financial consequences relating to legal and bank fees resulting from a dispute over my inheritance if I die prematurely.		
I understand that, no matter how difficult it is to contemplate, selling my farm when I retire may produce the best financial result for the family as a whole.		
Legal advice and documents	**Yes**	**No**
I've had legal advice about what will happen to the farm if I die suddenly.		
I've had legal advice about what would happen to the farm if one or more of my family members disputes my will and succession plan.		
I understand what probate is and how it is achieved.		
I have a will lodged with a solicitor.		
I've appointed a trusted friend or family member as an executor to my will.		
I understand the problems that will ensue if I die without a written will and have set a date with my solicitor to lodge one.		
I've established an enduring power of attorney enabling someone else to manage my affairs in the event of my incapacitation.		
The business structure of my farm is clear and well documented, as are any associated legal entities such as companies and trusts.		
The documents associated with my succession plan are co-located off-farm and a trusted friend or member of my family knows where to find them.		
Plans	**Yes**	**No**
I have a comprehensive farm succession plan and a trusted member of my family knows how to access it.		
I have a farm operational plan and a trusted member of my family knows how to access it.		
Details of my superannuation accounts and beneficiaries are documented and a trusted member of my family knows how to access this information.		

Continued

Appendix 1E. *Continued*

Farm budget and accounts	Yes	No
I update my farm budget on a monthly basis.		
My farm budget and accounts are easy to understand, and a trusted member of my family knows how to access them.		
My farm budget includes a list of assets and liabilities that I update on a quarterly basis.		
I've clearly documented the ownership status of any off-farm assets and a trusted member of my family knows how to access the documents.		

Appendix 1F: Improving your farm's environmental sustainability

Improving the environmental sustainability of your farm may seem daunting, but if you have a plan that starts with learning exactly what biological and physical resources you have on your farm, then you're well on your way.

Knowledge and experience	Yes	No
I know if I don't take action to maintain or improve the quality of my farm's environmental resources it will degrade environmentally and ultimately result in financial loss.		
I know that improving the quality of my farm's environmental resources will bring other benefits including improving the farm's appearance, my psychological wellbeing and broader social good.		
I'm open-minded about adopting new practices and seeing what does and doesn't work for environmental improvement.		
I intend to improve my general understanding of environmental science.		
I know where to get information about the environmental and physical resources on my land, such as native plants, animals and soil analysis.		
I've been researching the biological and physical resources on my land.		
I have some understanding of water flow and catchment dynamics.		
I can identify the major tree, shrub and grass species on my land.		
I can identify when and where the biota on my land is stressed or flourishing.		
I know what biodiversity is and how it can help my farming operations.		
I know what environmental services are available and how they can help my farming operations.		
I keep a seasonal diary that notes key biological events such as tree flowering, bird migrations and changes in pest animal numbers.		
Planning	**Yes**	**No**
I've calculated the optimum sustainable stocking rate of the grazing land on my property which preserves or improves its environment.		
I've assessed the environmental condition of my farm and what needs to be done to maintain or improve it.		
I have a 10-year environment plan that lays out projects, timeframes and costs.		
My 10-year plan includes planting an optimal number of tree plots.		
My 10-year plan includes improvements to waterways, dams and swamps.		
My 10-year plan includes a strategy for combating soil problems such as erosion, compaction, salinity and nutrient depletion.		
My 10-year plan includes a strategy for tackling pest animals and noxious weed species.		
Profitability	**Yes**	**No**
I know that a well-maintained property, with obvious environmental improvements, can sell for up to 25% more than a neglected farm.		

Continued

Appendix 1F. *Continued*

	Yes	No
I know that natural regenerative pasture techniques reduce the need for pesticides, herbicides and fertilisers, with associated cost savings.		
I've researched ways to improve the environment on my farm in a way that improves its profitability.		
I've done some cost–benefit analyses on various environment-improvement projects to work out which will give me the best return for my expenditure.		
I've visited other farms to study how environmental works have contributed to improved profitability.		
I've received advice from state and federal government agencies about how environmental betterment on my farm can improve its profitability.		
I've researched what government grants, subsidies and tax concessions might be available to help with environmental betterment on my farm.		
External agencies and community	**Yes**	**No**
I understand what government environmental legislation applies to land on my farm (e.g. the *Environment Protection and Biodiversity Conservation Act 1999*).		
I understand what federal and state environmental regulations apply to land on my farm.		
I know who to contact in government bodies to help me understand the biological and physical resources on my land.		
I've received information from government agencies about the biological and physical resources on my land.		
I've joined an industry-specific, farm sustainability project group.		
I know which NGOs help farmers with environmental works on their farms.		
I'm working with one or more NGOs to improve the environmental quality of my farm.		
I'm working with local community organisations to improve the environmental quality of my farm.		
My neighbours and I are helping each other improve the environmental quality of our farms.		
I've joined my local Landcare group.		
I've involved the local community in helping me plant trees.		
I've involved the local school in helping me plant trees.		
On-farm action	**Yes**	**No**
I've started small – cleaning up my farm, getting rid of rubbish and working out a full environmental improvement program.		
I've started planting trees.		
I've fenced off some riparian zones to protect them from stock.		
I've fenced off some areas of high biodiversity.		
I've started some regular, scientifically based monitoring projects and I'm recording the results to assess progress.		
I've started monitoring water quality in waterways and dams.		

Appendix 1G: Working with government

It is worth knowing as much as you can about how government agencies operate, what legislation and regulations you must comply with and what financial benefits you might be entitled to.

Personal and taxation benefits	Yes	No
I know which government agencies (e.g. Centrelink) provide personal and family benefits, and the circumstances under which they provide them.		
I know what tax benefits my family and I are eligible for under Australian taxation law.		
I know the difference between tax deductions and depreciation.		
My accountant is/I am knowledgeable and experienced with taxation legislation and know to claim whatever deductions and depreciation allowances I'm legally entitled to.		
I know what the federal Farm Management Deposits Scheme is and how I can access its benefits.		
I know the difference between income tax and capitals gains tax and how they can affect the sale value of my assets.		
Environmental legislation	**Yes**	**No**
I'm familiar with the federal *Environment Protection and Biodiversity Conservation Act 1999* and the implications for environmental and other works on my land.		
I'm familiar with state environmental regulations and how they affect what I can do on my property (e.g. in relation to land clearing).		
I've researched what federal and state government subsidies are available for environmental works, either directly or through third parties such as Greening Australia.		
I know what weeds and animal pests are listed as noxious under federal and state legislation and what I'm obliged to do to eradicate them.		
Animal welfare legislation	**Yes**	**No**
I'm familiar with animal welfare legislation relating to the treatment and disposal of livestock.		
I'm familiar with animal welfare legislation relating to the treatment and disposal of pest species.		
I understand that all native animals are protected species under conservation legislation.		
I understand how to dispose humanely of injured or dead livestock after adverse events such as fires, floods or cyclones.		
I understand my legal rights and responsibilities relating to trespassers who come onto my land, including activists who target my farming operations.		
Guns	**Yes**	**No**
I understand my rights and responsibilities in the purchase, safe use, transportation and storage of guns and ammunition.		

Continued

Appendix 1G. *Continued*

Development regulations	Yes	No
I understand what development regulations apply to my farm.		
Industry-specific regulations	**Yes**	**No**
I understand what industry-specific regulations apply to my farm.		
Farming support	**Yes**	**No**
I know which state and federal government departments to approach for information about various aspects of farming including agriculture in general, industry-specific issues, animal welfare, environment, commodities, trade and exports.		

Appendix 2: Adverse-event plans

Appendix 2A: Bushfire adverse-event plan

A severe bushfire will disrupt a farming business for several years. Rebuilding farm infrastructure and fences can take much longer than usual, especially if the entire area is burned and labour and supplies are in high demand locally, thus pushing up prices.

Your bushfire plan should evaluate the risk of fires happening in your region, and their possible intensity and severity.[1] It should document what action you will take in three categories – preparedness, event action and recovery. It should also include a time schedule for when each activity is to be completed, and costings where relevant.

Pre-event risk assessment and preparedness

Seek information from your state bushfire agency about bushfire readiness and preparing a fire survival plan.[2] Your plan should document what you need to do to be as well-prepared as possible for a fire. The list includes:

- interpreting weather data and its implications
- keeping up to date with real-time weather data (bookmark weather warning and emergency services websites)
- mapping fire-prone areas in terms of degree of risk, both within and around your property (e.g. in surrounding state forests and national parks)
- regularly backing up your farm computer files to an external system
- keeping all farm gates operable and fire trails free of debris
- creating fire breaks along outer perimeter fences, especially along public roads
- creating barriers to protect vital infrastructure (e.g. cyclone fencing around the homestead and major buildings)
- planting fire-resistant trees and shrubs in a break on the windward side of the homestead and other major infrastructure

- installing a roof-watering system
- installing an alternative power supply and fuel in a fireproof facility (the regional power supply may be cut off well before the fire reaches your property and you will need a generator to run water pumps and for essential household power needs such as refrigeration)
- having a large enough supply of tank water to use for firefighting with coupling that allows immediate connection to fire-fighting equipment
- having fire-fighting equipment, including water tank, pump, hoses and knapsacks
- having a chainsaw and knowing how to use it safely
- having an emergency kit of first aid supplies, and a list of contact details of key personnel
- knowing how you will dispose of injured and dead livestock
- knowing whether you are fully equipped to stay and defend your property, or will need to evacuate early
- having a fireproof safe space to retreat to should you not be able to contain the fire
- knowing how best to deal with a fire-front to save your property
- having a survival plan for yourself, your family, pets and service animals (e.g. horses and working dogs) and knowing the safest route out of your property.

Event action

Your plan should document the various actions you must take when you hear of a fire-front threatening your property. These include:

- checking your insurance policy is up to date
- ensuring valuable personal items and identification documents are either in a fireproof safe space or removed from the property
- checking the next 24 h temperature and wind forecast
- checking stock location and moving the animals to safer ground
- moving vehicles and machinery to open paddocks, sufficiently far apart that if one vehicle catches fire it will not readily set the others on fire
- opening gates wherever possible so that stock can escape from one paddock to another
- turning off mains power and gas
- locking your homestead and machinery buildings if you are evacuating
- keeping family, pets and working dogs close by, or evacuating them early
- letting your neighbours know whether you are staying or evacuating (note: while you might intend to stay, be prepared for the possibility that you will be evacuated by emergency services).

Post-event recovery

Your fire plan should outline how you will carry out the following post-event recovery activities:

- taking photographs before you start cleaning up so that you have a record for insurance purposes
- destroying suffering livestock (who will do it, how they will do it) and the disposal of carcasses
- demolishing and removing irreparably damaged infrastructure
- repairing fire-damaged farm buildings, yards and fences
- monitoring weed growth in pastures
- seeking personal and financial support from relevant agencies.

Phoenix farmer

A farmer, W.M. Waters of Jingellic in NSW, was burned out four times and rose from the ashes in each instance. He left the University of New England at 20 to help his father rebuild their family farm after a fire in 1972. Across his farming life he was burned out three more times but each time was able to re-establish his farming business, demonstrating that people can and do recover from fire. Mr Waters died in the cattle yards in late January 2020, aged 65. We should all strive to reach his extraordinary level of mental strength and resilience, and be open to seeking help when we need it to achieve this.

Endnotes

1   For example, a fire in a broadacre cropping region (e.g. the wheat-fields of Western Australia) will be of lower intensity than on a property that is heavily timbered.
2   In NSW, for example, see www.rfs.nsw.gov.au/plan-and-prepare/bush-fire-survival-plan

Appendix 2B: Flood adverse-event plan

Widespread catastrophic flooding is relatively uncommon in Australia, as are flash floods such as the one that struck Canberra in 1971 that resulted in the death of seven people in an urban area. Many rivers in flood-prone zones have been dammed in recent decades and their water-flow regulated; however, unexpected floods still occur and can cause considerable damage to farmland, infrastructure and stock, especially in association with cyclones (e.g. the 2018 Broome flood affected an area in the Kimberley the size of Victoria).

Your plan should evaluate the risk of a flood happening in your locality, and document what action you would take in three categories – risk assessment and preparedness, event action and recovery. It should also include a time schedule for when each activity will be completed, and costings where relevant.

Pre-event risk assessment and preparedness

Your plan should document the various things you need to consider in order to be as well-prepared as possible for a flood. These include:

- interpreting weather data and its implications
- keeping up to date with real-time weather data (bookmark weather warning and emergency services websites)
- mapping flood-prone areas in terms of degree of risk
- backing up your farm computer files to an external system
- creating diversion structures to protect vital infrastructure from floodwaters
- investigating refuge mounds for stock, including building small fenced areas for bulls, rams and stallions
- planting trees and shrubs that control erosion and slow the movement of flowing water
- installing an alternative power supply and fuel in a watertight facility or above the potential high-water mark
- storing two weeks of non-perishable food supplies and sufficient clean drinking water
- creating an emergency kit of first aid supplies and a list of contact details of key personnel
- storing a boat, outboard motor and oars somewhere handy
- knowing how you will dispose of dead livestock and flood debris
- having a family survival plan and knowing the safest route off your property.

Event action

Your plan should document the various actions you must take when you hear news of imminent floodwater arrival. These include:

- checking that your insurance policy is up to date
- storing or tying down all movable objects
- anchoring fuel tanks and other large objects that might float away
- moving waste, chemicals and poisons to high ground
- storing electrical tools in the highest possible place
- sandbagging doors and inside toilets, baths and handbasins to prevent sewage backflow
- ensuring that valuable personal items and identification documents are in a waterproof container
- checking stock location and moving animals to higher ground
- moving vehicles and machinery to higher ground
- opening gates wherever possible so that water can flow freely
- closing valves on irrigation systems
- within the homestead, moving small items that are at floor level onto furniture or upstairs, lifting rugs, etc.
- turning off water, power and gas
- locking your homestead and machinery buildings if you are evacuating
- keeping family, pets and working dogs close by, or evacuating them early
- letting your neighbours know whether you intend to stay or evacuate.

Post-event recovery

Waiting for floodwaters to subside can be trying. If you have been evacuated by emergency personnel, you may not be able to return until they give an all-clear signal. Avoid floodwaters because they might be contaminated by oil, gas, chemicals or raw sewage; boil tap water until you know it is safe to drink.

Your flood plan should state how you will complete the following post-event activities:

- taking photographs before you start cleaning up so that you have a record for insurance purposes
- disposing of dead livestock
- cleaning up homestead, farm buildings, yards and fences
- regrading roads
- monitoring weed growth in pasture
- seeking personal and financial support from relevant agencies.

Appendix 2C: Cyclone adverse-event plan

Cyclones can occur anywhere in the northern half of Australia, although most form off the north-west coast of Western Australia, in the Gulf of Carpentaria or in the Coral Sea. However, in some instances, damage can occur well south of landfall as the cyclone tracks inland on a southward trajectory. The cyclone season is November to April, and on average eight tropical cyclones and four severe tropical cyclones hit the Australian continent each year.

As well as strong winds, cyclones bring floods and storm surges. Because tropical cyclones form at sea and are monitored by the Bureau of Meteorology, you are likely to have at least 24 h warning before a cyclone's landfall, during which time its intensity can increase or decrease. Your plan should evaluate the risk of cyclones occurring in your region, and document what action you will take in three categories – risk assessment and preparedness, event action and recovery. It should also include a time schedule for when each activity will be completed, and costings where relevant.

Pre-event risk assessment and preparedness

Your plan should document the various things you have to do to be as well-prepared as possible for a cyclone. These include:

- interpreting weather data and its implications
- keeping up to date with real-time weather data (bookmark weather warning and emergency services websites)
- backing up your farm computer files to an external system
- constructing infrastructure to cyclone standard (where financially feasible)
- installing cyclone shutters on all windows, or installing cyclone-resistant windows
- planting cyclone-resistant tree windbreaks
- where allowed by state legislation, removing trees and branches close to buildings and other key infrastructure
- taking crop-specific action (e.g. trellising exotic tropical fruit trees)
- having back-up power generators and enough fuel to support essential activities (e.g. milking cows) and back-up communications systems
- having tarpaulins of sufficient size and number to cover livestock sheds should the roofing be lost
- having sufficient timber sheeting to board up windows and doors if storm shutters have not been installed
- having a strategy for the removal of debris and a site for dumping green waste afterwards
- having a family survival plan and knowing evacuation points.

Event action

Your plan should document the various actions you must take when you hear news of an imminent cyclone. These include:

- checking your insurance policy is up to date
- storing or tying down all movable objects
- installing window and door covers
- harvesting fruit early where possible
- sending stock to market where possible
- taking crop-specific action (e.g. removing the canopy of fruit trees to limit damage)
- creating temporary flood barriers
- keeping family, pets and working animals close by
- moving to evacuation points early
- letting your neighbours know whether you intend to stay or evacuate.

Post-event recovery

Your cyclone plan should state how you will carry out the following post-event activities (if flooding has occurred, then items in section 3 of the Flood Adverse-Event Plan will also be relevant). These include:

- taking photographs before you start cleaning up so that you have a record for insurance purposes
- disposing of dead livestock
- removing vegetation debris
- clearing and regrading roads
- cleaning up the homestead, farm buildings, yards and fences
- seeking personal and financial support from relevant agencies
- contacting industry support groups.

Appendix 3: Is a new property worth buying?

This appendix contains calculations associated with determining whether a particular property is worth buying (see Chapter 4) by examining levels of capital gains tax, net present value and loan repayments.

Capital gains tax calculation

Based on a sale price of $2 500 000 minus selling costs of $100 000 = $2 400 000
Capital outlay of $1 120 000 plus development costs of $120 000 = a net capital gain of $1 280 000
Divided by two for 50% capitals gains tax discount = $640 000 × 0.47 (the top tax rate) = $300 800 tax payable.
Net proceeds = $2 400 000 − $300 800 = $2 099 200 expected value

Net present value calculation on expected value

Capital: $2 099 200 × $(1+r)^n$ = $2 099 200 × $(1+0.08)^{-15}$ = $661 755
Income: 15 years back to Year 4 then discount back to Year 0

$$= \$25\,000 \times \frac{1-(1+r)^{-n}}{r} = \$25\,000 \times 7.1389 = \$178\,474$$

Discount back to Year 0: $178 474 × $(1+r)^{-n}$

$$= \$178\,474 \times (1+0.08)^{-4}$$

$$= \$178\,474 \times 0.735 = \$131\,181$$

Plus NPV of sale price $661 755 = $792 935.
Minus initial investment (purchase price plus development costs)

$$= \$792\,935 - \$1\,120\,000$$

$$= \text{NPV of } -\$327\,065.$$

This is a negative NPV, therefore it is not acceptable as an investment project. It would be better to look at the returns of some sort of off-farm investment instead.

Amortised loan calculations

A 15-year amortised loan (principal plus interest) with a principal of $500 000 at 8% interest (a 50% loan to value ratio):

$$= \$500\,000 \times \frac{r(1+r)^n}{(1+r)^n - 1}$$

$$= \$500\,000 \times (0.2537/2.1721)$$

$$= \$500\,000 \times 0.01168$$

$$= \$53\,399 \text{ annual debt service (interest compounded annually)}$$

$$= \text{debt service ratio of } 11.68\%$$

Assume net cash flow of $25 000 minus debt service of $58 399 = $33 399 cash deficit.

The farmer may increase debt service to $80 000 in good years to build a buffer against drought ('hay in the shed').

There is a large negative interest arbitrage: 8% minus 2.23%.

This is evaluating investment returns based on expected future value of farm expansion.

Stress-testing using the high and low variance could also be analysed by NPV.

References

There is an abundance of online information about all the subjects that are covered in this book. As well as online material, the following economics and farm finance reference books were employed in various places in the text to assist with analyses.

Anderson D (2014) *Endurance: Australian Stories of Drought*. CSIRO Publishing, Melbourne.

Australian Farm Institute (2003) *Farming Profits and the Environment*. Occasional Paper, June, p. 1. www.farminstitute.org.au/publications/occasional-papers.html; http://www.farminstitute.org.au/publications/occasional-papers.html

Australian Taxation Office Farm Management Deposits Scheme. www.business.gov.au/Grants-and-Programs/Farm-Management-Deposits-Scheme#:~:text=The%20Farm%20Management%20Deposits%20(FMD,of%20their%20risk%2Dmanagement%20strategy

Botsman R and Rogers R (2011) *What's Mine is Yours: The Rise in Collaborative Capitalism*. Harper Collins, Sydney.

Campbell H and Brown R (2003) *Benefit–Cost Analysis: Financial and Economic Appraisal Using Spreadsheets*. Cambridge University Press, Melbourne.

Castle EM, Becker MH and Smith FJ (1971) *Farm Business Management: The Decision-Making Process*. Macmillan, New York.

Chisholm AH and Dillon JL (1966) *Discounting and Other Interest Rate Procedures in Farm Management*. University of New England Press, Armidale.

Dairy Australia (online) *Our Farm, Our Plan* initiative. https://www.dairyaustralia.com.au/farm-business/our-farm-our-plan

Ehrenberg RG and Smith RS (2006) *Modern Labour Economics: Theory and Public Policy* (9th edn). Pearson Education, London.

Farm Table (online) *Farm Succession Planning: An Introduction and Helpful Guide*. https://farmtable.com.au/farm-succession-planning-information/

Keohane ND and Olmstead SM (2007) *Markets and the Environment*. Island Press, Washington (particularly Chapter 2, Economic efficiency and environment protection).

Lindenmayer DB, Michael D, Crane M, Florance D and Burns E (2018) *Restoring Farm Woodlands for Wildlife*. CSIRO Publishing, Melbourne.

Lockie S (2015) *Australia's Agricultural Future: The Social and Political Context*. Report to the Australian Council of Learned Academies. https://acola.org.au/wp/PDF/SAF07/social%20and%20political%20context.pdf

McRobert K (21 February 2019) *Briefing: Dairy Regulation and Floor Pricing.* Australian Farm Institute, Sydney. https://www.farminstitute.org.au/briefing-dairy-regulation-and-floor-pricing/.

Mickleboro J (2020) 4 high-yield ASX dividend shares for income investors. *The Motley Fool.* www.fool.com.au/2020/01/03/4-high-yield-asx-dividend-shares-for-income-investors/

Miles CA, Lockwood M, Walpole S and Buckley E (1998) Report No. 107. Charles Sturt University, Albury. In Australian Farm Institute (2003) *Farming Profits and the Environment.* Occasional Paper, June. Australian Farm Institute, Sydney.

Murphy D (30 July 2011) Shrunken industry fleeced by politics and greed. *Sydney Morning Herald.* www.smh.com.au/entertainment/books/shrunken-industry-fleeced-by-politics-and-greed-20110729-1i4at.html

Peirson G and Bird RG (1976) *Business Finance.* McGraw-Hill Australia, Sydney (particularly the financial tables at the back of the book).

Pyhrr SA and Cooper JR (1982) *Real Estate Investment: Strategy, Analysis, Decisions.* Wiley, Boston.

Rural Industries R&D Corporation Industry Overview (online) *Focus on Cyclone Resilience Research and Development.* Publication No. 13/122. www.agrifutures.com.au/wp-content/uploads/publications/13-122.pdf

Walpole S, Lockwood M and Miles CA (1998) *Influence of Remnant Native Vegetation on Property Sale Price.* Report No. 106. Johnstone Centre, Charles Sturt University, Albury.

Index

accountants 9, 12, 64, 65, 67, 68, 774
active income 34
adverse-event plans 9–13, 97–103
    bushfires 97–9
    drought 11–13
    elements 10
    off-farm income 10
    'self-insurance fund' 9–10
adverse-event shock 86–7
Agbiz tools (Queensland Government) 2, 3
    budgets 23–4, 25
aggregate cash-flow budget 27–8, 29
agricultural price stabilisation schemes 52–3
amortised loan calculations (buying a new
    property) 105
amortised loan repayments over 20 years 64
animal welfare legislation 95
annual net profit 30, 31
ANU Sustainable Farms project and website
    8, 72, 75
ANZ New Zealand *Rural Tools and Templates*
    5–6
assets list 28–30

bank/bank manager, relationship with 9, 12,
    75
biodiversity improvements 40, 71
    benefits of 75
    and law of diminishing returns 44
budgets 21–36
    Agbiz tools (Queensland Government)
        23–4, 25
    aggregate cash-flow budget 27–8
    categories 23
    creating 23–4

debt-reduction budget 24–5
development budget 24
environment improvement budget 25–6
household budget 25
importance of 21–2
maintenance budget 24
bushfires
    adverse-event plan 97–9
    adverse-event shock 86–7
business income 34
business plan *see* farm business plans
buying neighbouring land
    affording loan repayments 63–5
    calculating your farm business equity 65
    doubling your profit but doubling your
        risk 59–60
    further financial considerations 62–3
    is the property worth buying? 60–2, 104–5
    see also expanding your farming
        operations
buying your first farm, checklist 81–3

capital 40, 44
capital expenditure projects, comparison of
    four alternative 51–2
capital gains tax calculation (buying a new
    property) 104
cash-flow budget 27–8, 29
cattle yards, new, expenditure evaluation 52
cleaning up your farm 75
commodity prices 64
    for export products 55
    fluctuations in rural 52–6
cost–benefit analysis 8, 38, 40, 41–3
    environmental improvements 72–3

example 42
increasing stocking rates (example) 45–6
investing in pasture improvement
(example) 47–50
and law of diminishing marginal returns
43–6
procedure 42
purpose 41–2
time value of money and net present value
46–52
CSIRO website 8
cyclones
adverse-event plan 102–3
adverse-event shock 86–7

dairy products two-price arrangement 53
debt accumulation, example 21–2
debt-reduction budget 24–5
debt servicing (repayments) 2–4, 32, 63–5
debt tunnels 4, 32
debts, reducing by selling burdensome assets
74
development budget 24
drought, adverse-event shock 86
'drought declared' areas 12
drought survival plan 11–13
how much water and feed do you have?
11–12
know your region 11
and off-farm investment strategy 33, 35
what are the financial risks? 12–13

environmental improvements
budget 25–6
cost–benefit analysis 72–3
intergenerational and financial health 73
and mental health 74, 75
environmental investment, improving farm
profits through 70–2
environmental legislation and regulations 94,
95
environmental plans 7–9
cost–benefit analysis 8
goals 7–8
sustainable farming practices 8–9
environmental remediation 82–3, 90
farmer benefits 71
farmer scepticism about benefits of 70

tension between farmers and governments
over 70–1
environmental sustainability
achieving 70–7
improving, checklist 93–4
inspiring examples 8–9
proven strategies for achieving 71–2
websites 8
equity see farm business equity
estate planning 15–16
executors 15
expanding your farming operations
checklist 88–90
is the property worth buying? 61–2, 104–5
optimising your farming operations 66–8
see also buying neighbouring land
external agencies 94

failure in farming 74
family cohesion 75
family farms
and succession 15–16, 17–18
succession dispute, case example 16–17
family relationships 2, 74, 83, 90
farm balance sheet, example 21–2
farm business equity 21–2, 30, 31
calculation 65
calculator 26
statement 2–3
farm business plans 4–19
adverse-event plans 9–13, 97–103
environmental plans 7–9
goals 5
importance of 4
insurance 13–14, 87
making plans 4–5
overarching business plans 5–7
recording achievements 7
reviews and updates 6
SMART maxim 5
software systems 6
succession plans 14–18, 91–2
templates 5–6
farm commodity prices 54
farm environmental sustainability see
environmental sustainability
farm expansion, financing 59–69
farm finance checklists 80–97

farm machinery *see* machinery
Farm Management Deposits Scheme 1, 9–10, 23, 33
farm profitability
    factors affecting 39
    improving, checklist 84–5
    increasing 38–58
    and optimal farm size 69
farmers
    financial challenges and decisions 78
    and governments' environmental objectives 70–1
    scepticism about benefits of environmental remediation 70
farm's assets and liabilities, list of 28–30
fencing 7, 8, 43, 70, 71, 75
financial health and intergenerational environmental improvement 73
financial shock 86
financial stress 21, 32, 35, 74
financing farm expansion 59–69
fire
    adverse-event plan 97–9
    adverse-event shock 86–7
fixed-term deposits 33
floods
    adverse-event plan 100–1
    adverse-event shock 86–7
foreclosure 4, 32
frugal, how to be 22–3

genetic improvements 40
glossary vii
government legislation and regulations 82, 83, 90, 94, 95, 96
governments
    environmental objectives, impact on farmers 70–1
    working with, checklist 95–6
guns 95

hiring tools 22
household budget 25

improving your farm's environmental sustainability, checklist 93–4
improving your farm's profitability 38–58
    checklist 84–5

income
    active and passive 34
    business 34
    off-farm 10, 33–4, 87
    off-farm investment 34–5
inflation 24, 27, 46, 55, 64
insurance 13–14, 87
    level of insurance 14
    policy types 13–14
    reimbursement following a claim 14
intergenerational environmental improvements and financial health 73
international currency exchange rates 55
investment income, off-farm 33, 34–5, 66
investment property, negative gearing 66
investment(s)
    off-farm 33–4
    value of a farm as 31
is the property worth buying? 61–2
    financial calculations 104–5

labour/labour costs 39, 40, 41, 44, 67
land values 64, 71
large farms 59
law of diminishing (marginal) returns 38–9
    and cost–benefit analysis 43–6
    examples 43, 44
    increasing stocking rates (example) 45–6
legislation 83, 90, 94, 95
liabilities list 30
list of farm's assets and liabilities 28–30
loan repayments 2–4, 24–5, 32
    buying additional land 63–5

machinery
    costs of 40, 44, 66, 67
    update, expenditure evaluation 52
maintenance budget 24
making plans 4–19
Meat and Livestock Australia website 8
mechanisation 40, 67
mental health 2, 74, 83, 90
    and environmental improvements 74, 75
    and wellbeing 74–5
millennium drought 11, 35
monopsony power, impact of 53–4
motivation, optimal level of 68–9

negative gearing 35, 66
net equity 2, 3, 30, 31
net present value (NPV)
    calculation on expected value (buying a
        new property) 104–5
    comparison of four alternative capital
        expenditure projects 51–2
    investing in pasture improvement
        (example) 47–51
    and time value of money 46–52
net profit 31
net worth (equity) statement 2–3
noxious weed control 7

off-farm employment and business
    income 34
off-farm income 10, 33–4, 87
off-farm investment income 33, 34–5, 66
optimising your farming operations 66
    optimal level of motivation 68–9
    optimal size of your operation 67
    optimal way of financing expansion 68
    optimal way of managing your operation
        67
overarching business plans 5–7

passive income 34
pasture improvement, investing in 47
    cost–benefit analysis 47–52
    financing analysis 50–1
pasture improvements, to boost productivity
    40
pest control 7
photographic records 7
productivity gains, and terms of trade 54–6
productivity improvements 40–1
projects 5, 8, 38, 44, 46
    see also capital expenditure projects
purchasers' market power 53, 54

rate of return on capital 30–1
real estate investment 9, 34–5, 64, 66
receivership 16
regulations 82, 83, 90, 94, 95
retirement/retirement plans 2, 15, 72
return on capital 30–1
return on investment from a farm 31
riparian areas 7, 25, 40, 71, 72

rural commodity price fluctuations 52–3
    impact of monopsony power 53–4
    terms of trade and productivity gains
        54–6

savings 87
    power of 32–3
seasonal conditions, and price changes 54
'self-insurance' fund 9
selling burdensome assets 74
selling your farm 2
shares 9, 35, 66
shearing shed, new, expenditure
    evaluation 52
shock-proofing your farm, checklist 86–7
SMART maxim 5
soil erosion reduction 7, 8
soil improvement 7
spreading risk 35
stocking rates, increasing 45–6
structural improvements 40
succession plans 14–18
    checklist 91–2
    dispute over a family farm, case example
        16–17
    documentation and executors 15
    and family members 15–17
    and off-farm investments 33
    online resources 14–15
    'tough love' recommendation 17–18, 33
    types of 15
    see also wills
sustainable farming practices 8–9, 70–7

taking stock 1–2
    mechanics of 2–4
tax minimisation 66
terms of trade and productivity gains 54–6
    calculations for the author's farm 55–6
time value of money 46
    and net present value 46–52
tools
    hiring vs buying 22–3
    safeguarding 23
tree plantings 7, 40, 44
    benefits of 75
    budget example 25
    direct subsidies 8, 75

expenditure evaluation 51–2
as uplifting experience 75

'useful life' of assets 28

value of a farm as an investment 31

water-distribution system 41

water efficiency 40
waterway restoration 7
expenditure evaluation 51–2
wellbeing 74–5
wills 15–16
Wool Reserve Price Scheme 53
working with government, checklist 95–6